Switching Circuits
For Engineers

PRENTICE-HALL ELECTRICAL ENGINEERING SERIES

Switching Circuits

For Engineers

Third Edition

MITCHELL P. MARCUS

Senior Engineer
IBM Corporation

Adjunct Assistant Professor
State University of New York at Binghamton

PRENTICE-HALL, INC., Englewood Cliffs, New Jersey

Library of Congress Cataloging in Publication Data

Marcus, Mitchell P
  Switching circuits for engineers.

  Includes bibliographical references.
  1. Switching theory. 2. Electric networks.
3. Logic circuits. I. Title.
TK7868.S9M3 1975   621.3815'37   74–13060
ISBN 0–13–879908–3

PRENTICE-HALL INTERNATIONAL, INC., *London*
PRENTICE-HALL OF AUSTRALIA, PTY. LTD., *Sydney*
PRENTICE-HALL OF CANADA, LTD., *Toronto*
PRENTICE-HALL OF INDIA PRIVATE LIMITED, *New Delhi*
PRENTICE-HALL OF JAPAN, INC., *Tokyo*

To *El*
Bunny
Glenn
Lee
and *Ricky*

Contents

Preface

Switching circuits are circuits that perform logic functions. The design of switching circuits is referred to as logic design. Switching circuits can range in complexity from a simple staircase lighting circuit, controlled from both upstairs and downstairs, to a complex circuit that performs arithmetic in an electronic digital computer.

The interesting task of the logic designer is to properly interconnect basic logic circuit elements or "logic blocks" so that the resultant circuit performs a desired logic function. There are usually many different ways in which these logic blocks may be interconnected to realize a desired function; however, some of these ways require more logic blocks than others. So the logic designer's task goes further than merely realizing the desired logic function; he tries, in general, to realize the function with the most economical circuit that he can.

The logic designer has at his disposal many formalized methods for designing and simplifying switching circuits. These methods can not only lead to simpler circuits, but can considerably reduce the time required to reach a solution. Furthermore, the elimination of even a few redundant circuit components can increase the reliability of circuit operation, reducing down-time and servicing; it can make the circuit easier to understand, simplifying training and trouble-shooting; and, if the circuit is involved in a high production unit, the elimination of only a few components can be multiplied into a substantial cost saving.

In addition to his work in switching circuit theory and its application to the logic design of IBM products, the author has been teaching courses in switching circuits at IBM since 1954. He has also served as Assistant Professor at the State University of New York at Binghamton, where he has taught graduate level courses in switching circuits.

This book is an outgrowth of this extensive involvement in both the practical usage and teaching of switching circuits. It is written for both the logic design engineer and the student in switching circuits. The aim has been to present the material in a way that it can be most easily understood and usefully applied. The book is recommended for a one-semester course at either the undergraduate or graduate level.

Boolean algebra is covered in Chapters 1 and 2. The algebra is not merely "presented" to the reader; the reader is shown how to apply the algebra in logic minimization and is given "methods of attack."

Chapter 3 discusses the relationship of the algebra to logic circuits; the fundamental concepts of basic and detailed logic diagramming; positive and negative logic; the relationship between the logic NOT function (logic complementation) and the electrical inverter circuit (electrical inversion); and the standard symbols.

Many non-logic factors may be involved in the conversion from a derived Boolean expression to the physical circuit, depending upon the "hardware" involved; these considerations are not within the scope of this book. The aim is to keep the approach general, and adaptable to any technology. Therefore, any discussion of our everchanging hardware has purposely been omitted; such information is readily available from many sources.

In Chapter 4, the concept of optional combinations is introduced, and the tabular and iterative consensus methods of minimization are presented, both for single and multiple output functions.

In Chapter 5, the map method of minimization is presented, along with a "method of attack"; both single and multiple output functions are considered.

In Chapter 6, an analysis of number systems is presented, and adders are discussed. Chapter 7 discusses codes, error detection and correction, and the related concept of minimum distance.

The treatment of sequential circuits has been completely revised: rather than a separate treatment for each type of sequential circuit e.g., pulse, clocked, level, etc, a *unified* approach has been developed.

Chapter 8 introduces sequential circuits. It includes an intuitive look at sequential circuit synthesis and presents an overview of the steps in the formal synthesis. A chapter is then devoted to each step.

Chapter 9 discusses state diagrams, state tables, and flow tables, and their construction from the word statements of the problems. Chapter 10 describes state and flow table reduction, and the concepts of equivalence, compatibility, and merging. Chapter 11 treats state assignment, and discusses races and cycles.

Chapter 12 deals with excitation maps and expressions. The various kinds of flip flops and other memory elements are thoroughly discussed. The universal map method is described; this is an original method in which the excitation expressions for any type of memory element can be read from a single map.

Chapter 13 completes the synthesis of sequential circuits with a discussion of output maps and expressions, and the problem of hazards. To recapitulate the synthesis procedure for sequential circuits, some illustrative examples are worked out in their entirety.

Appendices on relay circuits follow. Relay circuits present an additional vehicle for the understanding of switching circuits. Appendix A relates Boolean expressions to contact networks. In Appendix B, symmetric functions are discussed, and their implementation in contact networks described. Appendix C discusses sequential relay circuits and hazards peculiar to these circuits.

For the reader who would like to delve deeper into the subject of switching circuits, an extensive list of related literature for further study is included. The list is arranged chronologically by chapter, to help the reader obtain a historical perspective of any selected phase of the subject.

Lengthy examples and problems, such as total design problems, tend to obscure the essential points under study. It is therefore generally advantageous from a learning standpoint for the problems to be built around a pertinent portion of a design. The examples and problems used in this book are, for the most part, purposely short and to the point, and confined to the topic under study.

Problems are presented at the end of the chapters to give the reader an opporutnity to test his knowledge and understanding of the subject. Answers to the majority of the problems are given at the back of the book; solutions are also included where it is felt that they would be helpful. Some of the end-of-chapter problems are identified by an asterisk; for these problems no answers are furnished at the back of the book; these problems can be used by instructors for assignments or for testing.

The author is indebted to the many IBM students and State University of New York at Binghamton students who gave the first two editions such a severe workout, and who, by their probing questions and comments, contributed to this drastically revised third edition.

M. P. MARCUS
Binghamton, New York

Switching Circuits
For Engineers

1

Boolean Algebra

Policy No. 22 may be issued only if the applicant

1. Has been issued Policy No. 19 and is a married male,
or 2. Has been issued Policy No. 19 and is married and under 25,
or 3. Has not been issued Policy No. 19 and is a married female,
or 4. Is a male under 25,
or 5. Is married and 25 or over.

From an XYZ Insurance Company Manual

Can you simplify the statement above? There is a great deal of redundancy in this policy statement. Using intuition only, most people will not be able to recognize all of the redundancy in as simple a statement as this one. An equivalent but simpler statement appears near the end of this chapter.

Boolean algebra, an algebra of logic, enables us to dispense with intuition and deductively simplify logic statements that are even much more complex. Boolean algebra is named after George Boole who, in the middle 1800s, developed it. Almost a hundred years later Claude E. Shannon realized its application to the simplification of logic circuits or switching circuits.

Experience has shown that if one learns Boolean algebra with relation to its circuit implications he does not become proficient in it because he "thinks"

1

in circuits rather than in the algebra. Experience has also shown that the study of Boolean algebra from a purely abstract point of view is not attractive to most engineers because there is no practical association for them to "hang their hat on." Study of Boolean algebra as it relates to logic statements has been found to be the most effective initial approach and this approach is followed here. Later, it will be shown how the algebra, one of the most basic tools available to the logic designer, can be used to simplify logic circuits.

$X = 1$ *or else* $X = 0$

Consider basic logic statements that must be either true or false. For example: *The applicant is a male.* Letter symbols are used to represent such statements as follows:

$$X = \text{the applicant is a male}$$

It can now be said that X must be true or else X must be false. Carrying our symbolism a step further, a 1 is used to represent the "value" of a true statement, and a 0 is used to represent the "value" of a false statement. If the statement "The applicant is a male" is true, we say that the "value" of X is 1, written $X = 1$. If the statement is false, we say that the "value" of X is 0, written $X = 0$.

Thus,

$$X = 1 \quad \text{or else} \quad X = 0$$

There is no numerical significance to the 1 and 0; there is only a logic significance.

Although 1 and 0 represent the truth or falsity of a statement, the prerogative is taken of saying "X equals 1" or "X equals 0."

AND

In the reduction of compound logic statements to Boolean algebra, there are three key words of special importance: AND, OR, and NOT. First consider a compound statement made up of two basic logic statements connected by the word AND.

The applicant is a male AND the applicant is married

Making use of symbology

$$X = \text{the applicant is a male}$$
$$Y = \text{the applicant is married}$$

1

Boolean Algebra

Policy No. 22 may be issued only if the applicant

   1. Has been issued Policy No. 19 and is a married male,
or 2. Has been issued Policy No. 19 and is married and under 25,
or 3. Has not been issued Policy No. 19 and is a married female,
or 4. Is a male under 25,
or 5. Is married and 25 or over.

From an XYZ Insurance Company Manual

Can you simplify the statement above? There is a great deal of redundancy in this policy statement. Using intuition only, most people will not be able to recognize all of the redundancy in as simple a statement as this one. An equivalent but simpler statement appears near the end of this chapter.

Boolean algebra, an algebra of logic, enables us to dispense with intuition and deductively simplify logic statements that are even much more complex. Boolean algebra is named after George Boole who, in the middle 1800s, developed it. Almost a hundred years later Claude E. Shannon realized its application to the simplification of logic circuits or switching circuits.

Experience has shown that if one learns Boolean algebra with relation to its circuit implications he does not become proficient in it because he "thinks"

in circuits rather than in the algebra. Experience has also shown that the study of Boolean algebra from a purely abstract point of view is not attractive to most engineers because there is no practical association for them to "hang their hat on." Study of Boolean algebra as it relates to logic statements has been found to be the most effective initial approach and this approach is followed here. Later, it will be shown how the algebra, one of the most basic tools available to the logic designer, can be used to simplify logic circuits.

X = 1 or *else X* = 0

Consider basic logic statements that must be either true or false. For example: *The applicant is a male*. Letter symbols are used to represent such statements as follows:

$$X = \text{the applicant is a male}$$

It can now be said that X must be true or else X must be false. Carrying our symbolism a step further, a 1 is used to represent the "value" of a true statement, and a 0 is used to represent the "value" of a false statement. If the statement "The applicant is a male" is true, we say that the "value" of X is 1, written $X = 1$. If the statement is false, we say that the "value" of X is 0, written $X = 0$.
Thus,

$$X = 1 \quad \text{or else} \quad X = 0$$

There is no numerical significance to the 1 and 0; there is only a logic significance.

Although 1 and 0 represent the truth or falsity of a statement, the prerogative is taken of saying "X equals 1" or "X equals 0."

AND

In the reduction of compound logic statements to Boolean algebra, there are three key words of special importance: AND, OR, and NOT. First consider a compound statement made up of two basic logic statements connected by the word AND.

The applicant is a male AND the applicant is married

Making use of symbology

$$X = \text{the applicant is a male}$$
$$Y = \text{the applicant is married}$$

1

Boolean Algebra

Policy No. 22 may be issued only if the applicant

    1. Has been issued Policy No. 19 and is a married male,
or 2. Has been issued Policy No. 19 and is married and under 25,
or 3. Has not been issued Policy No. 19 and is a married female,
or 4. Is a male under 25,
or 5. Is married and 25 or over.

From an XYZ Insurance Company Manual

Can you simplify the statement above? There is a great deal of redundancy in this policy statement. Using intuition only, most people will not be able to recognize all of the redundancy in as simple a statement as this one. An equivalent but simpler statement appears near the end of this chapter.

Boolean algebra, an algebra of logic, enables us to dispense with intuition and deductively simplify logic statements that are even much more complex. Boolean algebra is named after George Boole who, in the middle 1800s, developed it. Almost a hundred years later Claude E. Shannon realized its application to the simplification of logic circuits or switching circuits.

Experience has shown that if one learns Boolean algebra with relation to its circuit implications he does not become proficient in it because he "thinks"

in circuits rather than in the algebra. Experience has also shown that the study of Boolean algebra from a purely abstract point of view is not attractive to most engineers because there is no practical association for them to "hang their hat on." Study of Boolean algebra as it relates to logic statements has been found to be the most effective initial approach and this approach is followed here. Later, it will be shown how the algebra, one of the most basic tools available to the logic designer, can be used to simplify logic circuits.

$X = 1$ *or else* $X = 0$

Consider basic logic statements that must be either true or false. For example: *The applicant is a male.* Letter symbols are used to represent such statements as follows:

$$X = \text{the applicant is a male}$$

It can now be said that X must be true or else X must be false. Carrying our symbolism a step further, a 1 is used to represent the "value" of a true statement, and a 0 is used to represent the "value" of a false statement. If the statement "The applicant is a male" is true, we say that the "value" of X is 1, written $X = 1$. If the statement is false, we say that the "value" of X is 0, written $X = 0$.

Thus,

$$X = 1 \quad \text{or else} \quad X = 0$$

There is no numerical significance to the 1 and 0; there is only a logic significance.

Although 1 and 0 represent the truth or falsity of a statement, the prerogative is taken of saying "X equals 1" or "X equals 0."

AND

In the reduction of compound logic statements to Boolean algebra, there are three key words of special importance: AND, OR, and NOT. First consider a compound statement made up of two basic logic statements connected by the word AND.

The applicant is a male AND the applicant is married

Making use of symbology

$$X = \text{the applicant is a male}$$
$$Y = \text{the applicant is married}$$

The entire compound statement can now be written as

$$X \text{ AND } Y$$

When is X AND Y true and when is X AND Y false? X may be true or false and Y may be true or false. Taken together, there are four possibilities: X and Y may both be true, X may be true and Y false, X may be false and Y true, or both X and Y may be false.

X AND Y is true only if X is true and Y is true. This can be tabulated as follows:

X		Y	X AND Y
True	AND	True	True
True	AND	False	False
False	AND	True	False
False	AND	False	False

A "·" is used to symbolize AND. Thus, X AND Y is written $X \cdot Y$. Replacing *True* with 1, *False* with 0, and AND with "·" gives the following relationships:

$$
\begin{array}{ccc}
X & Y & X \cdot Y \\
\hline
1 \cdot 1 & = & 1 \\
1 \cdot 0 & = & 0 \\
0 \cdot 1 & = & 0 \\
0 \cdot 0 & = & 0 \\
\end{array}
$$

Although the "·" signifies multiplication in ordinary algebra, here it has only the logic AND significance. Other symbols have also been used in the literature to represent AND; among them are $\wedge$ and $\cap$.

OR

Now consider a compound statement made up of two basic logic statements connected by the word OR.

The applicant is a male OR the applicant is married

Substituting X and Y for the two statements, as before, gives

$$X \text{ OR } Y$$

This OR is the *inclusive* OR; that is, X OR Y means X or Y or both. (With the *exclusive* OR, X OR Y would mean either X or Y but not both.) Unless specifically stated otherwise, OR will always be understood to mean the *inclusive* OR.

When is X OR Y true and when is X OR Y false? The same four possible combinations of X and Y exist. X OR Y is true when X is true or when Y is true or when both are true. This is tabulated as follows:

X		Y	X OR Y
True	OR	True	True
True	OR	False	True
False	OR	True	True
False	OR	False	False

A "$+$" is used to symbolize OR. Thus, X OR Y is written $X + Y$. Replacing *True* with 1, *False* with 0, and OR with "$+$" gives the following relationships:

$$
\begin{array}{ccc}
X & Y & X + Y \\
\hline
1 + 1 = & & 1 \\
1 + 0 = & & 1 \\
0 + 1 = & & 1 \\
0 + 0 = & & 0 \\
\end{array}
$$

Although the "$+$" signifies addition in ordinary algebra, here it has only the logic OR significance. Some other symbols that have been used to represent OR are $\vee$ and $\cup$.

NOT

Now consider the statement

The applicant is NOT a male

When is this statement true and when is it false? If the statement "The applicant is a male" is true, then the statement "The applicant is NOT a male" is false. If the statement "The applicant is a male" is false, then the statement "The applicant is NOT a male" is true. This can be tabulated very simply by letting X represent the statement "The applicant is a male" and letting NOT X represent the statement "The applicant is NOT a male":

X	NOT X
True	False
False	True

Many different symbols have been used to symbolize NOT. For instance, NOT X can be written as $\bar{X}$, X', $1 - X$, or $\sim X$. We shall use the symbol $\bar{X}$ to represent NOT X. There is less chance of "losing" the "bar" than of losing the "prime," and the bar can be applied to an expression without the need of adding parentheses; that is, $\overline{X + Y}$ will be used instead of $(X + Y)'$ or $1 - (X + Y)$ or $\sim(X + Y)$.

If a statement is true only when a second statement is false, and vice versa, as in the case above, the two statements are said to be *complements* or *negations* of each other. Thus, $\bar{X}$ is the complement of X, and X is the complement of $\bar{X}$.

Using our symbology,

$$\text{if} \quad X = 1, \quad \text{then} \quad \bar{X} = \bar{1} = 0$$

$$\text{if} \quad X = 0, \quad \text{then} \quad \bar{X} = \bar{0} = 1$$

This is summarized in the following table:

X	$\bar{X}$
1	0
0	1

We now have, by definition,

$$\bar{1} = 0 \quad \text{and} \quad \bar{0} = 1$$

It also follows that

$$\bar{\bar{1}} = 1$$
$$\bar{\bar{0}} = 0$$
$$\bar{\bar{X}} = X$$

Postulates

Following is a summary of the relationships so far:

$X = 1$ *or else* $X = 0$	
$1 \cdot 1 = 1$ $1 \cdot 0 = 0 \cdot 1 = 0$ $0 \cdot 0 = 0$	$0 + 0 = 0$ $0 + 1 = 1 + 0 = 1$ $1 + 1 = 1$
$\bar{1} = 0$	$\bar{0} = 1$

This summary represents the *postulates* of Boolean algebra. Based on these postulates are many useful theorems that enable us to manipulate and simplify logic expressions.

Boolean algebra, like any other algebra, is composed of a set of symbols and a set of rules for manipulating these symbols. However, some differences between ordinary algebra and Boolean algebra should be stressed here. In ordinary algebra the letter symbols may take on a large or even an infinite number of values; in Boolean algebra they may assume only one of two possible values, 0 and 1. Thus, Boolean algebra is much simpler than ordinary algebra. In ordinary algebra the values have a *numerical* significance; in Boolean algebra, they have a *logic* significance. Furthermore, the meanings of "·" and "+" in Boolean algebra—AND and OR—are entirely unrelated to their meanings in ordinary algebra—"times" and "plus."

In one sense the choice of "·," "+," 1, and 0 is unfortunate because of the tendency to associate them with their counterparts in ordinary algebra. In another sense, the choice is advantageous because of the coincidental relationship that five of the six postulates involving the "·" and "+" bear to their meanings in ordinary algebra.

Some Definitions

The different letters in a Boolean expression are called *variables*. For example, in the expression

$$A \cdot \bar{B} + \bar{A} \cdot C + A \cdot (D + E)$$

there are five variables, A, B, C, D, and E. Each occurrence of a variable or its complement is called a *literal*. In the expression above there are seven literals. The "·" is usually omitted in writing expressions in Boolean algebra and is implied merely by writing the literals, or factors, in juxtaposition. Thus,

$$A \cdot \bar{B} + \bar{A} \cdot C + A \cdot (D + E)$$

would normally be written as

$$A\bar{B} + \bar{A}C + A(D + E)$$

The "·" is used only where additional clarity is required.

Two expressions are *equivalent* if one expression equals 1 only when the other equals 1, and one equals 0 only when the other equals 0. Two expressions are *complements* of each other if one expression equals 1 only when the other equals 0, and vice versa.

The complement of a Boolean expression is obtained by

changing all ·'s to +'s

changing all +'s to ·'s

changing all 1's to 0's

changing all 0's to 1's

and complementing each literal

Thus, the complement of

$$1 \cdot A + \bar{B}C + 0$$

is

$$(0 + \bar{A})(B + \bar{C}) \cdot 1$$

When the first expression equals 1, the second equals 0, and vice versa.

The *dual* of a Boolean expression is obtained by

changing all ·'s to +'s

changing all +'s to ·'s

changing all 1's to 0's

changing all 0's to 1's

but not complementing any literal

Thus, the dual of

$$1 \cdot A + \bar{B}C + 0$$

is

$$(0 + A)(\bar{B} + C) \cdot 1$$

There is no general relationship between the "values" of dual expressions; that is, both may equal 1, both equal 0, or one may equal 1 while the other equals 0. Duals are of principal interest in the study of the Boolean postulates and theorems, and are also useful in simplification procedures, as we shall see later.

In the preceding table of postulates, the six postulates involving the "·" and "+" have been purposely arranged in three rows of two postulates each. Each pair of postulates may be considered as either complements or duals of each other since no literals are involved. The theorems that follow are presented in dual pairs.

Theorems

Many useful theorems derived from the postulates, will now be studied. These theorems enable us to simplify logic expressions or transform them into other useful equivalent expressions.

1a. $0 \cdot X = 0$ **1b.** $1 + X = 1$

In ordinary algebra it is not generally possible to prove a theorem by substituting all possible values of the variables since there may be a large or an infinite number of values. In Boolean algebra, since the variables can have only two values, 0 and 1, theorems can easily be proved merely by testing their validity for all possible combinations of values of the variables involved. This type of proof is called proof by perfect induction.

Theorem 1a may be proved as follows: X must equal either 0 or 1. If $X = 0$, then $0 \cdot 0 = 0$. If $X = 1$, then $0 \cdot 1 = 0$. Thus, no matter what the value of X,

$$0 \cdot X = 0$$

Theorem 1b can be proved in an analogous manner. However, the proof can be approached differently by first writing the theorem so that it is in complementary form to Theorem 1a. The theorem in this form would read $1 + \bar{X} = 1$. Based on the fact that every postulate has a complementary postulate, if a theorem is valid, then its complementary theorem is valid. This is so because if a theorem is true, based on certain postulates, then its complementary theorem must be true based on the complementary postulates. Thus, Theorem 1a having been proved, the complementary theorem $1 + \bar{X} = 1$ must also be true.

Since the validity of a theorem is based on its being true for all possible combinations of values of the variables, there is no reason the $\bar{X}$ in the theorem cannot be replaced by an X. Thus, if the theorem $1 + \bar{X} = 1$ holds for all values of $\bar{X}$, the theorem $1 + X = 1$ must be true for all values of X. Therefore, inconsequential "bars" over the variables are omitted in the theorems. The expression $1 + X = 1$ is the dual of the expression $0 \cdot X = 0$. Thus, if a theorem is valid, then its dual theorem must also be valid. For this reason, it is not necessary to go through the mechanics of proving both of a pair of dual theorems to be true; proving one is sufficient.

The literals in a theorem may represent not only single variables but also terms or longer expressions. For example, using Theorem 1,

$$0 \cdot (A\bar{B} + C) = 0 \qquad 1 + A\bar{B} + C = 1$$

The important point to remember about Theorem 1 is that

$$0 \cdot anything = 0$$

and

$$1 + anything = 1$$

2a. $1 \cdot X = X$ $\qquad\qquad$ **2b.** $0 + X = X$

This pair of theorems can be proved as easily as the first pair by substituting both possible values for X. The important point to remember about Theorem 2 is that

multiplication by 1

or

addition of 0

does not affect an expression. For example,

$$1\cdot(A\bar{B} + C) = A\bar{B} + C \qquad 0 + A\bar{B} + C = A\bar{B} + C$$

The prerogative is taken of using the terms "multiplication" and "addition" to represent the AND $(\cdot)$ operation and OR $(+)$ operation, respectively. The term "product" is used to represent the result of the AND operation. Thus, XYZ is called the product of X, Y, and Z; $(A + B)(C + D)$ is called the product of $A + B$ and $C + D$. The term "sum" is used to represent the result of the OR operation. Thus, $X + Y + Z$ is called the sum of X, Y, and Z; $AB + CD$ is called the sum of AB and CD. Furthermore, an expression such as $(A + B)(C + D)$ is called a "product of sums," and $(A + B)$ and $(C + D)$ are called "sum terms." An expression such as $AB + CD$ is called a "sum of products," and AB and CD are called "product terms."

3a. $XX = X$ **3b.** $X + X = X$

An example of the application of Theorem 3 follows:

$$(A\bar{B}\bar{B} + C)(A\bar{B} + C + C) = (A\bar{B} + C)(A\bar{B} + C) = A\bar{B} + C$$

This example stresses again that the literals in these theorems may represent not only single variables but also more complex expressions.

4a. $X\bar{X} = 0$ **4b.** $X + \bar{X} = 1$

If $X = 1$, then $\bar{X} = 0$; if $X = 0$, then $\bar{X} = 1$. In either case, Theorem 4a represents the product of 1 and 0, which is 0, whereas Theorem 4b represents the sum of 1 and 0, which is 1. Theorem 4 says that

anything multiplied by its complement $= 0$

and

anything added to its complement $= 1$

Some simple exercises on Theorems 1 through 4 follow. These exercises should be tried before reading the solutions that follow.

Simplify:

(a) AAB (b) $A\bar{A}B$
(c) $A\bar{A} + B$ (d) $A + A + B$
(e) $A + \bar{A} + B$ (f) $(A + \bar{A})B$

Solutions:

(a) $AAB = AB$

(b) $A\bar{A}B = 0 \cdot B = 0$

(c) $A\bar{A} + B = 0 + B = B$

(d) $A + A + B = A + B$

(e) $A + \bar{A} + B = 1 + B = 1$

(f) $(A + \bar{A})B = 1 \cdot B = B$

5a. $XY = YX$

5b. $X + Y = Y + X$

EXAMPLE

$$A\bar{B} + C = \bar{B}A + C = C + A\bar{B} = C + \bar{B}A$$

6a. $XYZ = X(YZ) = (XY)Z$

6b. $X + Y + Z = X + (Y + Z)$
$$= (X + Y) + Z$$

7a. $\overline{XY \ldots Z} = \bar{X} + \bar{Y} + \ldots + \bar{Z}$ **7b.** $\overline{X + Y + \ldots + Z} = \bar{X}\bar{Y} \ldots \bar{Z}$

This theorem is known as DeMorgan's theorem. Theorem 7a can be proved as follows: If $X, Y \ldots$, and Z all equal 1,

$$\overline{1 \cdot 1 \cdot \ldots \cdot 1} = \bar{1} + \bar{1} + \ldots + \bar{1}$$
$$\bar{1} = \bar{1}$$
$$0 = 0$$

If $X, Y \ldots$, and Z do not all equal 1, then one or more of these literals must equal 0. If even one of the literals equals 0,

$$\overline{0 \cdot 1 \cdot \ldots \cdot 1} = \bar{0} + \bar{1} + \ldots + \bar{1}$$
$$\bar{0} = 1 + 0 + \ldots + 0$$
$$1 = 1$$

Theorem 7a states that a product of literals may be complemented by changing the product to a sum of the literals and complementing each literal. Theorem 7b states that a sum of literals may be complemented by changing the sum to a product of the literals and complementing each literal.

EXAMPLES

$$\overline{\bar{A}BC\bar{D}E} = A + \bar{B} + \bar{C} + \dot{D} + \bar{E}$$

$$\overline{A + \bar{B} + \bar{C} + D + \bar{E}} = \bar{A}BC\bar{D}E$$

Note that Theorems 7a and 7b and the above examples represent equivalences. For example, $\overline{\bar{A}BC\bar{D}E}$ is *equivalent* to $A + \bar{B} + \bar{C} + D + \bar{E}$, but $\bar{A}BC\bar{D}E$ and $A + \bar{B} + \bar{C} + D + \bar{E}$ are *complements* of each other.

This pair of theorems may be written in a more general form as in Theorem 8.

8. $$\bar{f}(X, Y, \ldots, Z, \cdot, +) = f(\bar{X}, \bar{Y}, \ldots, \bar{Z}, +, \cdot)$$

This theorem is read as follows. Given: an expression containing literals such as X, Y, and Z, and occurrences of the operators $\cdot$ and $+$. To complement this expression, signified by the $\bar{f}$, each literal is complemented, each $\cdot$ is changed to $+$, and each $+$ is changed to $\cdot$. A simple example follows.

$$\overline{C + A\bar{B}} = \bar{C}(\bar{A} + B)$$

Note the importance of the parentheses. In the original expression, C is added to the product $A\bar{B}$. Therefore, the complement $\bar{C}$ must be multiplied by the sum $(\bar{A} + B)$. Now for a more complex example:

$$\overline{(A\bar{B} + C)\bar{D} + E} = [(\bar{A} + B)\bar{C} + D]\bar{E}$$

Again note the importance of parentheses and brackets. The product $A\bar{B}$, when complemented, becomes $(\bar{A} + B)$. This product was originally added to C; therefore, the sum $(\bar{A} + B)$ is now multiplied by $\bar{C}$. The sum $(A\bar{B} + C)$ was originally multiplied by $\bar{D}$; therefore, the product $(\bar{A} + B)\bar{C}$ is now added to D. Finally, E was originally added to $(A\bar{B} + C)\bar{D}$; therefore, $\bar{E}$ is now multiplied by $(\bar{A} + B)\bar{C} + D$, giving $[(\bar{A} + B)\bar{C} + D]\bar{E}$.

Some important simplification theorems now follow.

9a. $XY + XZ = X(Y + Z)$ **9b.** $(X + Y)(X + Z) = X + YZ$

Theorem 9a is like factoring in ordinary algebra. The operation represented by Theorem 9b is not permitted in ordinary algebra, but the procedure is analogous to that in Theorem 9a. The procedure may be better understood if Theorem 9a is first considered in a little different way. The X is common to both terms XY and XZ. X is *multiplied* by Y, and X is *multiplied* by Z. Therefore, X will be *multiplied* by the *sum* of the remainders of each term, namely $(Y + Z)$, giving $X(Y + Z)$.

Now, Theorem 9b can be thought of in a similar way. X is common to both terms $(X + Y)$ and $(X + Z)$. X is *added* to Y, and X is *added* to Z. Therefore, X will be *added* to the *product* of the remainders of each term, namely YZ, giving $X + YZ$.

EXAMPLES

(a) $$AB + A\bar{C}D + A(E + \bar{F}) = A(B + \bar{C}D + E + \bar{F})$$
(b) $$(A + B)(A + \bar{C} + D)(A + E\bar{F}) = A + B(\bar{C} + D)E\bar{F}$$

The examples have purposely been presented in dual pairs so that the similarity of the dual operations can be more easily seen. This practice will be maintained throughout the study of theorems.

In example (a), A is common to all three terms. In each case A is *multiplied* by the remainder of the term. Therefore, the A will be *multiplied* by the *sum* of what remains in each term: A is multiplied by the sum $(B + \bar{C}D + E + \bar{F})$. $E + \bar{F}$ does not require additional parentheses (see Theorem 6).

In example (b), A is common to every term. In each case, A is *added* to the remainder of the term. Therefore, the A will be *added* to the *product* of what remains in each term: A is added to $B(\bar{C} + D)E\bar{F}$. Again, $E\bar{F}$ does not require additional parentheses (see Theorem 6).

Now for two slightly more involved examples:

(a) $$A\bar{B}CD + A\bar{B}CE + ACF = AC(\bar{B}D + \bar{B}E + F)$$
$$= AC[\bar{B}(D + E) + F]$$

(b) $$(A + \bar{B} + C + D)(A + \bar{B} + C + E)(A + C + F)$$
$$= A + C + (\bar{B} + D)(\bar{B} + E)F$$
$$= A + C + (\bar{B} + DE)F$$

In example (a), AC is common to all three terms and is *multiplied* by the remainder of a term in each case. Therefore, AC is *multiplied* by the *sum* of the remainder of each term, namely $(\bar{B}D + \bar{B}E + F)$. Furthermore, within the parentheses, $\bar{B}$ is common to two terms; therefore, $\bar{B}$ is similarly factored out to obtain the final expression.

In example (b), $A + C$ is common to all three terms and is *added* to the remainder of the term in each case. Therefore, $A + C$ is *added* to the *product* of the remainder of each term, namely $(\bar{B} + D)(\bar{B} + E)F$. Furthermore, $\bar{B}$ is common to two of the terms, leading to the final expression.

10a. $XY + X\bar{Y} = X$ **10b.** $(X + Y)(X + \bar{Y}) = X$

This theorem may appear to be a special case of Theorem 9. In Theorem 10a,

$$XY + X\bar{Y} = X(Y + \bar{Y})$$
$$= X \cdot 1 = X$$

In Theorem 10b,

$$(X + Y)(X + \bar{Y}) = X + Y\bar{Y}$$
$$= X + 0 = X$$

However, this theorem has further implications. In a sum or product of 2^m n-variable terms, if m variables occur in all possible combinations (represented by Y and $\bar{Y}$ in the theorem), while the remaining $n - m$ variables are constant (represented by X in the theorem), the m variables are redundant and the $n - m$ variables define the expression.

An example with $n = 3$ and $m = 2$ is as follows:

$$X\bar{Y}\bar{Z} + X\bar{Y}Z + XY\bar{Z} + XYZ = X \cdot 1 = X$$

Here, in $2^m = 2^2 = 4$ terms, $m = 2$ variables (Y and Z) occur in all possible combinations, while $n - m = 1$ variable (X) is constant; thus, Y and Z are redundant and X defines the expression.

Another example with $n = 4$ and $m = 2$ is as follows:

$$(\bar{W} + X + \bar{Y} + \bar{Z})(\bar{W} + X + \bar{Y} + Z)$$
$$(\bar{W} + X + Y + \bar{Z})(\bar{W} + X + Y + Z) = \bar{W} + X + 0 = \bar{W} + X$$

Y and Z occur in all possible combinations, whereas $\bar{W}$ and X are constant; thus, the expression reduces to $\bar{W} + X$.

Note that the number of terms involved in such a simplification must be a power of two (2, 4, 8, 16, etc.), since there are 2^m combinations of m variables.

This theorem is the basis of other simplification methods which will be taken up in Chapters 4 and 5.

11a. $X + XY = X$ **11b.** $X(X + Y) = X$

Theorem 11a can be proved as follows:

$$X + XY = X(1 + Y) = X \cdot 1 = X$$

Although Theorem 11b can be considered proved once Theorem 11a is proved, since the theorems are duals of each other, Theorem 11b can also be proved as follows:

$$X(X + Y) = XX + XY = X + XY$$

and the rest of the steps will be the same as in the proof of Theorem 11a. Another proof is as follows:

$$X(X + Y) = (X + 0)(X + Y) = X + 0 \cdot Y = X + 0 = X$$

This simplification theorem may be applied in the following way. If a smaller term appears in a larger term, then the larger term is redundant. The smaller term is defined as the one containing fewer literals; conversely, the larger term is the one containing more literals. The larger is said to *subsume* the smaller.

In Theorem 11a, X appears in the larger term XY. Therefore, the term XY is redundant and the expression reduces to X. Similarly, in Theorem 11b,

X appears in the larger term $(X + Y)$; therefore, the term $(X + Y)$ is redundant and the expression reduces to X.

EXAMPLES

(a) $$A\bar{B} + A\bar{B}C + A\bar{B}(D + E) = A\bar{B}$$

(b) $$(A + \bar{B})(A + \bar{B} + C)(A + \bar{B} + DE) = A + \bar{B}$$

In example (a), the first term $A\bar{B}$ appears in the second and third terms; therefore, the second and third terms are redundant and the expression reduces to $A\bar{B}$. In example (b), the first term $(A + \bar{B})$ appears in the second and third terms; the second and third terms are therefore redundant and this expression reduces to $A + \bar{B}$.

12a. $X + \bar{X}Y = X + Y$ **12b.** $X(\bar{X} + Y) = XY$

A few interesting ways to prove Theorem 12a follow:

$$\begin{aligned}
X + \bar{X}Y &= (X + \bar{X})(X + Y) \quad \text{(Theorem 9b in reverse)} \\
&= 1 \cdot (X + Y) \\
&= X + Y
\end{aligned}$$

or

$$\begin{aligned}
X + \bar{X}Y &= X + XY + \bar{X}Y \quad \text{(Theorem 11a in reverse)} \\
&= X + Y \quad\quad\quad\quad\quad \text{(Theorem 10a)}
\end{aligned}$$

Theorem 12b can be proved in the same manner.

Proofs of the type just shown represent interesting manipulations of the algebra. However, a straightforward method of proof that can always be used is called the "truth table" proof, which is a means of applying the method of perfect induction. In a proof by perfect induction, it is shown that an equality of expressions exists for all possible combinations of values of the variables. This type of proof is especially adaptable to Boolean algebra where the variables can have only two values, 0 or 1. A truth table proof of Theorem 12a follows:

1 X	2 Y	3 $\bar{X}$	4 $\bar{X}Y$	5 $X + \bar{X}Y$	6 $X + Y$
0	0	1	0	0	0
0	1	1	1	1	1
1	0	0	0	1	1
1	1	0	0	1	1

First, every possible combination of the values of the variables is listed. In this case, with two variables, X and Y, there are four possible combinations, 00, 01, 10, and 11. These combinations are listed in columns 1 and 2. Since an $\bar{X}$ will be needed, the complementary values of column 1 are written in column 3. Next, the product $\bar{X}Y$ is required; this is placed in column 4, and is obtained by the multiplication of columns 2 and 3. In column 5 is written the values of the sum $X + \bar{X}Y$, which is obtained by the addition of columns 1 and 4. Finally, in column 6 is written the sum $X + Y$, which is obtained by the addition of columns 1 and 2. It is now found that the values in columns 5 and 6 agree for every possible combination of the variables X and Y, thus proving the theorem.

In practice it is found helpful to think about Theorem 12 in a slightly more general way, as shown in Theorem 12′.

12a′. $ZX + Z\bar{X}Y = ZX + ZY$
12b′. $(Z + X)(Z + \bar{X} + Y) = (Z + X)(Z + Y)$

The reason for presenting this modification of Theorem 12 is that the application is usually encountered in the form of Theorem 12′. The way of applying this theorem is as follows: If a smaller term appears in a larger term except that *one* variable in the smaller term and the corresponding variable in the larger term are complements, then that variable in the *larger* term is redundant.

In Theorem 12a′, the smaller term ZX appears in the larger term $Z\bar{X}Y$, except for the complementary X and $\bar{X}$. Therefore, the $\bar{X}$ in the larger term is redundant, and the expression reduces to $ZX + ZY$. A similar relationship exists in Theorem 12b′. It does not matter where the "bar" actually appears; it is always the variable in the *larger* term that is redundant.

EXAMPLE

$$W\bar{X} + WXY = W\bar{X} + WY$$

The X in the larger term is redundant because it is the complement of the $\bar{X}$ in the smaller term.

Note the difference between the application of Theorems 11 and 12′. Theorem 11: If a smaller term appears in a larger term with *no* complementations, *the entire larger term* is redundant. Theorem 12′: If a smaller term appears in a larger term except for *one* complemented variable, *only that variable in the larger term* is redundant. If a smaller term appears in a larger term with two or more variables complemented, no simplification of this sort is possible.

Now for some examples involving Theorems 11 and 12′.

(a) $A\bar{B} + A\bar{B}C + \bar{A}\bar{B}D + ABE + \bar{A}BF = A\bar{B} + \bar{B}D + AE + \bar{A}BF$

(b) $(A + \bar{B})(A + \bar{B} + C)(\bar{A} + \bar{B} + D)(A + B + E)(\bar{A} + B + F)$

$$= (A + \bar{B})(\bar{B} + D)(A + E)(\bar{A} + B + F)$$

In example (*a*), the first term appears in the second with no complementation; therefore, the entire second term is redundant. In each of the third and fourth terms, one variable is complemented: $\bar{A}$ in the third, and B in the fourth; these two variables are therefore redundant. In the fifth term, two variables are complemented and therefore no simplification of this sort is possible. The final expression may be factored in one of two possible ways if desired. Example (b) is the dual of example (a) and the same reasoning can be made throughout.

The following pair of theorems can be thought of as the "included term" theorems.

13a. $XY + \bar{X}Z + YZ = XY + \bar{X}Z$
13b. $(X + Y)(\bar{X} + Z)(Y + Z) = (X + Y)(\bar{X} + Z)$

An interesting proof of Theorem 13a is as follows:

$$\begin{aligned} XY + \bar{X}Z + YZ &= XY + \bar{X}Z + YZ(X + \bar{X}) \\ &= XY + \bar{X}Z + XYZ + \bar{X}YZ \\ &= XY + \bar{X}Z \end{aligned}$$

Theorem 13b may be proved in a similar manner.

Following is a truth table proof of Theorem 13a.

X Y Z	$\bar{X}$	XY	$\bar{X}Z$	YZ	$XY + \bar{X}Z$	$XY + \bar{X}Z + YZ$
0 0 0	1	0	0	0	0	0
0 0 1	1	0	1	0	1	1
0 1 0	1	0	0	0	0	0
0 1 1	1	0	1	1	1	1
1 0 0	0	0	0	0	0	0
1 0 1	0	0	0	0	0	0
1 1 0	0	1	0	0	1	1
1 1 1	0	1	0	1	1	1

The last two columns have the same value for all possible combinations of values of the variables, proving the equivalence. However, more can be learned from this truth table. Examination of the YZ column and the $XY + \bar{X}Z$ column will show that $XY + \bar{X}Z$ equals 1 for four of the eight possible combinations, whereas YZ equals 1 for two of the eight possible combinations. Furthermore, the two combinations for which YZ equals 1 are included among the four combinations for which $XY + \bar{X}Z$ equals 1; that is, the expression $XY + \bar{X}Z$ "includes" the term YZ—hence the name included term theorem.

Now for the recognition of the application of this pair of theorems. In the application of Theorem 13a, two terms are looked for: one that contains a variable, and the other that contains the complement of this same variable. For instance, the first term in the theorem contains an X and the second term contains an $\bar{X}$. If two such terms are found, the remainders of each term, exclusive of this variable and its complement, together form a product that is included by the first two terms. In Theorem 13a, the first term contains an X and the second term contains an $\bar{X}$. The remainders of these two terms are Y and Z, respectively, and together they form a product YZ which is included by the first two terms. An included term may lead to the elimination of redundancy in the expression. For example, in the theorem, the third term YZ is redundant: There is no need to add the term YZ to the terms $XY + \bar{X}Z$, since the term YZ is already logically included.

In a sense, Theorem 13a is first applied in reverse to obtain the included term. The included term may then be used to eliminate redundancy in the expression, following which Theorem 13a is applied to eliminate the included term.

Similar reasoning applies in Theorem 13b. Two terms are looked for: one that contains a variable, and the other that contains the complement of this variable. If two such terms are found, the remainders of each term, exclusive of this variable and its complement, together form a sum that is included by the first two terms. An included term may be used to eliminate redundancy in the expression.

Theorem 13 can often be used in conjunction with other theorems, such as Theorems 10, 11, and 12'. Some examples follow:

(a)
$$AB + \bar{A}C + BCD = AB + \bar{A}C$$

The first two terms, noting the A and $\bar{A}$, include a term BC. Because of the included term BC, the term BCD is redundant (Theorem 11a). Therefore the expression reduces to $AB + \bar{A}C$.

(b)
$$(A + B)(\bar{A} + C)(B + C + D) = (A + B)(\bar{A} + C)$$

The first two terms, again noting the A and $\bar{A}$, include a term $(B + C)$. $(B + C)$ appears in the third term $(B + C + D)$; therefore, the term $(B + C + D)$ is redundant (Theorem 11b).

(c)
$$AB + \bar{A}C + \bar{B}CD = AB + \bar{A}C + CD$$
$$= AB + C(\bar{A} + D)$$

The first two terms include the term BC. BC appears in the third term $\bar{B}CD$ except that the $\bar{B}$ in the third term is complemented. Therefore, the $\bar{B}$ is redundant (Theorem 12a') and the expression reduces to $AB + \bar{A}C + CD$ (which may be factored).

The next example is the dual of example (c).

(d) $\quad (A + B)(\bar{A} + C)(\bar{B} + C + D) = (A + B)(\bar{A} + C)(C + D)$
$$= (A + B)(C + \bar{A}D)$$

(e) $\qquad\qquad AB + \bar{A}C + \bar{B}C = AB + C$

The first two terms include BC. BC and $\bar{B}C$ reduce to C (Theorem 10a). The expression at this point reads $AB + \bar{A}C + C$. The $\bar{A}C$ term is redundant because of the C term. Therefore, the expression reduces to $AB + C$.

The next example is the dual of example (e).

(f) $\qquad\qquad (A + B)(\bar{A} + C)(\bar{B} + C) = (A + B)C$

(g) $\qquad\qquad ABC + \bar{A}BD + BCDE = ABC + \bar{A}BD$

The first two terms include BCD, which makes $BCDE$ redundant.

Included terms may lead to other included terms. For example,

(h) $\qquad\qquad AB + \bar{A}C + \bar{B}D + CD = AB + \bar{A}C + \bar{B}D$

The first two terms include BC. BC and $\bar{B}D$ include CD. Therefore, the fourth term CD is redundant, and the expression reduces to $AB + \bar{A}C + \bar{B}D$.

Additional examples are given for analysis.

(i) $\quad (A + B)(\bar{A} + C)(\bar{B} + D)(C + D) = (A + B)(\bar{A} + C)(\bar{B} + D)$

(j) $\quad AB + \bar{A}C + \bar{B}D + \bar{D}E + CE = AB + \bar{A}C + \bar{B}D + \bar{D}E$

or

$$AB + \bar{A}C + \bar{B}D + \bar{D}E + CE = AB + \bar{A}C + \bar{B}D + \bar{D}E$$

(k) $(A + B)(\bar{A} + C)(\bar{B} + D)(\bar{D} + E)(C + E)$
$$= (A + B)(\bar{A} + C)(\bar{B} + D)(\bar{D} + E)$$

(l) $AB + \bar{A}C + A\bar{C} = \bar{A}C + A\bar{C} + BC$

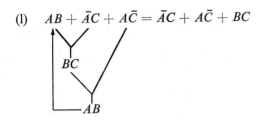

If the term AB is eliminated, the term BC cannot be eliminated since it would no longer be included by $AB + \bar{A}C$.

Resumé of Simplification Theorems and "Method of Attack"

While no hard and fast rules can be given for the best "method of attack" in simplifying any Boolean expression, the following approach is given as a guide.

Resumé of Simplification Theorems

1a. $0 \cdot X = 0$ **1b.** $1 + X = 1$
2a. $1 \cdot X = X$ **2b.** $0 + X = X$
3a. $XX = X$ **3b.** $X + X = X$
4a. $X\bar{X} = 0$ **4b.** $X + \bar{X} = 1$

. . .

9a. $XY + XZ = X(Y + Z)$ **9b.** $(X + Y)(X + Z) = X + YZ$
10a. $XY + X\bar{Y} = X$ **10b.** $(X + Y)(X + \bar{Y}) = X$
11a. $X + XY = X$ **11b.** $X(X + Y) = X$
12a. $X + \bar{X}Y = X + Y$ **12b.** $X(\bar{X} + Y) = XY$
12a'. $ZX + Z\bar{X}Y = ZX + ZY$
12b'. $(Z + X)(Z + \bar{X} + Y) = (Z + X)(Z + Y)$
13a. $XY + \bar{X}Z + YZ = XY + \bar{X}Z$
13b. $(X + Y)(\bar{X} + Z)(Y + Z) = (X + Y)(\bar{X} + Z)$

Theorems 1 to 4 are, of course, applied whenever possible; however, their application becomes almost "second nature," and more deliberate thought is usually directed toward the other less obvious theorems. It is stressed again that X may represent not only a variable but also a term or more complex expression.

Theorems 10, 11, and 12′ should be applied exhaustively. Then Theorem 13 should be applied. Theorems 10 through 12′ may further be applied in conjunction with or following the application of Theorem 13.

If a "factored" form, rather than a sum of products or product of sums form, is desired, then Theorem 9 may be applied. Theorem 9 generally should not be applied until there is no longer any possible application of the other theorems; if Theorem 9 is applied too early, the application of the other theorems may be obscured.

There is a tendency for the beginner to prefer to work with the sum of products form, rather than the product of sums form. To this end he may (1) "multiply out" a product of sums expression (that is, apply Theorem 9a in reverse) to obtain an equivalent sum of products expression; or (2) complement the product of sums expression to obtain a complementary sum of products, and after simplification, recomplement; or (3) obtain the dual of the product of sums expression and, after simplifying, obtain the dual of the dual.

This is an undesirable practice! Every additional operation adds a potential source of error. Also, multiplying out generally adds additional redundancy which must be removed. There is no need for this practice: Each sum of products theorem has its dual product of sums theorem, and both can be used with equal facility.

Additional Theorems

The pair of theorems that follow are not simplification theorems but rather transposition theorems.

14a. $XY + \bar{X}Z = (X + Z)(\bar{X} + Y)$
14b. $(X + Y)(\bar{X} + Z) = XZ + \bar{X}Y$

The key point to look for in the possibility of making a transposition of this type is two terms, one that contains a variable, and the other that contains the complement of this same variable. In Theorem 14a, the first term contains an X and the second term contains an $\bar{X}$; therefore, the transposition shown can be made by adding the X to the remainder of the term containing the $\bar{X}$, and adding the $\bar{X}$ to the remainder of the term containing the X, these two sums being multiplied together. Conversely, in Theorem 14b there are two terms, one that contains an X, and the other that contains an $\bar{X}$. The transposition shown can be made by multiplying the X by the remainder of the term containing the $\bar{X}$, and multiplying the $\bar{X}$ by the remainder of the term containing the X, and then adding these two products together.

EXAMPLE

$$A\bar{B}C + \bar{A}(\bar{D} + E) = (A + \bar{D} + E)(\bar{A} + \bar{B}C)$$

In making the transposition, the A is added to the remainder of the term containing the $\bar{A}$, namely $\bar{D} + E$, and the $\bar{A}$ is added to the remainder of the term containing the A, namely $\bar{B}C$, the two sums being multiplied together.

For a second example, the transposition will be made in the other direction, starting with $(A + \bar{D} + E)(\bar{A} + \bar{B}C)$. Here the A is multiplied by the remainder of the term containing the $\bar{A}$, namely $\bar{B}C$, and the $\bar{A}$ is multiplied by the remainder of the term containing the A, namely $\bar{D} + E$, the two products being added together. The result is the original expression $A\bar{B}C + \bar{A}(\bar{D} + E)$.

When the relationship between Boolean algebra and switching circuits is taken up later, the desirability of such transpositions will be apparent. Two special cases of the application of this theorem are given because of the frequency with which they are encountered in practice:

$$X\bar{Y} + \bar{X}Y = (X + Y)(\bar{X} + \bar{Y})$$
$$\bar{X}\bar{Y} + XY = (\bar{X} + Y)(X + \bar{Y})$$

The first case describes the "exclusive or" function (one or the other but not both), sometimes symbolized by $X \oplus Y$ or $X \vee\!\!\!\vee Y$. Another useful form of the expression is $(X + Y)\overline{XY}$. The second case describes the complement of the "exclusive or": the "neither or both" or "if and only if" function, sometimes symbolized by $X \equiv Y$. Another useful form of the expression is $\overline{(X + Y)} + XY$.

15a. $X \cdot f(X, \bar{X}, Y, \ldots, Z) = X \cdot f(1, 0, Y, \ldots, Z)$
15b. $X + f(X, \bar{X}, Y, \ldots, Z) = X + f(0, 1, Y, \ldots, Z)$

Theorem 15a states that if a variable X is multiplied by an expression containing occurrences of X or $\bar{X}$, then all X's in the expression may be replaced by 1's, and all $\bar{X}$'s in the expression may be replaced by 0's. This can be seen to be permissible since

$$X \cdot X = X \cdot 1 = X$$

and

$$X \cdot \bar{X} = X \cdot 0 = 0$$

Theorem 15b states that if a variable X is added to an expression containing occurrences of X and $\bar{X}$, then all X's in the expression may be replaced by 0's, and all $\bar{X}$'s may be replaced by 1's. Again this is permissible since

$$X + X = X + 0 = X$$

and

$$X + \bar{X} = X + 1 = 1$$

EXAMPLES

(a) $A \cdot [AB + \bar{A}C + (A + D)(\bar{A} + E)] = A \cdot [1 \cdot B + 0 \cdot C + (1 + D)(0 + E)]$
$$= A[B + 0 + 1 \cdot E]$$
$$= A(B + E)$$

(b) $\bar{A} \cdot (AB + \bar{A}C + D) = \bar{A} \cdot (0 \cdot B + 1 \cdot C + D)$
$$= \bar{A}(C + D)$$

Not only is this theorem useful for simplification, but it is partly the basis of the following theorem.

16a. $f(X, \bar{X}, Y, \ldots, Z) = X \cdot f(1, 0, Y, \ldots, Z) + \bar{X} \cdot f(0, 1, Y, \ldots, Z)$
16b. $f(X, \bar{X}, Y, \ldots, Z) = [X + f(0, 1, Y, \ldots, Z)][\bar{X} + f(1, 0, Y, \ldots, Z)]$

This pair of theorems can be proved using Theorem 10 in reverse, along with Theorem 15. In Theorem 16a an expression is multiplied first by X and also by $\bar{X}$, the two products being added together:
$$f(X, \bar{X}, Y, \ldots, Z) = X \cdot f(X, \bar{X}, Y, \ldots, Z) + \bar{X} \cdot f(X, \bar{X}, Y, \ldots, Z)$$
It can be seen that these two expressions are equivalent since the latter can be reduced to the former by the application of Theorem 10.

Now, by the application of Theorem 15 the X's and $\bar{X}$'s can be replaced with 1's and 0's, respectively, when the expression is multiplied by X, and they can be replaced by 0's and 1's, respectively, when the expression is multiplied by $\bar{X}$. Theorem 16b can be similarly proved.

Theorem 16 has the following application: Given an expression containing any number of occurrences of some variable and its complement, say X and $\bar{X}$, the expression can be rewritten using only one occurrence of X and one occurrence of $\bar{X}$, at most.

EXAMPLE

$$AB + \bar{A}C + (A + D)E + (\bar{A} + F)G$$

Find an equivalent expression with only one occurrence of A and one occurrence of $\bar{A}$, at most.

$$AB + \bar{A}C + (A + D)E + (\bar{A} + F)G$$
$$= A[AB + \bar{A}C + (A + D)E + (\bar{A} + F)G]$$
$$+ \bar{A}[AB + \bar{A}C + (A + D)E + (\bar{A} + F)G]$$

$$= A[1 \cdot B + 0 \cdot C + (1 + D)E + (0 + F)G]$$
$$+ \bar{A}[0 \cdot B + 1 \cdot C + (0 + D)E + (1 + F)G]$$
$$= A[B + 0 + 1 \cdot E + FG] + \bar{A}[0 + C + DE + 1 \cdot G]$$
$$= A[B + E + FG] + \bar{A}[C + DE + G]$$

The application of Theorem 16b is analogous.

While this pair of theorems reduces one variable to one occurrence of itself and its complement at most, it may introduce multiplicity of other variables. Circuit requirements, however, may make this a desirable operation.

Theorem 16 may be further applied to each bracketed expression independently, thereby reducing a selected second variable to two occurrences of itself and its complement, at most. A third selected variable may be reduced to four occurrences of itself and its complement, etc.

Boolean algebra will now be applied to the simplification of the *XYZ* Insurance Company Manual statement at the beginning of this chapter.

Let A = Applicant has been issued Policy No. 19

            B = Applicant is married

            C = Applicant is a male

            D = Applicant is under 25

Policy No. 22 may be issued only if

     1. ABC
or   2. ABD
or   3. $\bar{A}B\bar{C}$
or   4. CD
or   5. $B\bar{D}$

which can be written

$$ABC + ABD + \bar{A}B\bar{C} + CD + B\bar{D}$$

$= ABC + AB \ \ \ + \bar{A}B\bar{C} + CD + B\bar{D}$	(Theorem 12a′)
$= \qquad\quad AB \ \ + \ \ B\bar{C} + CD + B\bar{D}$	(Theorems 11a, 12a′)
$= \qquad\quad AB \ \ + \ \ B \ \ + CD + B\bar{D}$	(Theorems 13a, 10a)
$= \qquad\qquad\qquad\quad B \ \ + CD$	(Theorem 11a)

Policy No. 22 may be issued only if the applicant

     1. Is married,
or   2. Is a male under 25.

PROBLEMS

1. Simplify:
   (a) $A + \bar{B} + \bar{A}B + (A + \bar{B})\bar{A}B$
   (b) $(A + \bar{B} + \bar{A}B)(A + \bar{B})\bar{A}B$
   (c) $A + \bar{B} + \bar{A}B + \bar{C}$
   (d) $(A + \bar{B} + \bar{A}B)\bar{C}$
   (e) $(A + \bar{B})\bar{A}B + \bar{C}$
   (f) $(A + \bar{B})\bar{A}B\bar{C}$
   Hint: $A + \bar{B}$ and $\bar{A}B$ are complements.

2. Complement:
   (a) $[(\bar{A}B + \bar{C})D + \bar{E}]F$
   (b) $S[\bar{W} + I(T + \bar{C})] + H$
   *(c) $F[\bar{R}(I + \bar{D}A) + \bar{Y}]$
   *(d) $U + [(V + \bar{W})X + \bar{Y}]Z$

3. Reduce to minimum number of literals:
   (a) $CD(E + A)F + (A + E)BC$
   (b) $(A + BF + C + E)(D + E + FB)$
   (c) $ABC(D + E) + F(E + D)(G + H)B$
   (d) $(A + CE + B + F)(D + GH + F + EC)$
   (e) $AB(C + D)(\bar{E} + F) + \bar{G}(D + C)\bar{H}A$
   (f) $(A\bar{B} + \bar{C} + D + EF)(\bar{G} + FE + \bar{C} + \bar{H}K)$
   *(g) $BC(\bar{D} + F)(G + \bar{H}) + J(F + \bar{D})\bar{K}B$
   *(h) $(L + \bar{M} + N\bar{P} + \bar{Q}R)(R\bar{Q} + S + \bar{M} + T)$
   *(i) $\bar{A}(B + C)\bar{D}E + EF(C + B)(\bar{G} + H)$
   *(j) $(F\bar{R} + I + D + AY)(YA + \bar{Q} + U + I)$

4. Simplify:
   (a) $A\bar{B}\bar{C} + A\bar{B}\bar{C}D + \bar{C}A$
   (b) $A\bar{B}C + \bar{A}\bar{C}D + \bar{C}A$
   (c) $(A + B + CD)(\bar{A} + B)(\bar{A} + B + E)$
   (d) $DEH + \bar{E}G\bar{H} + \bar{H}E + HF\bar{E} + J\bar{H}E$
   (e) $(K + L + \bar{P})(L + M + P)(Q + P + \bar{L})(L + \bar{P})(\bar{P} + N + \bar{L})$
   (f) $(A + BC)(A + \bar{B} + \bar{C} + D)(\bar{A} + BC + E)(\bar{A} + \bar{B} + \bar{C} + F)$
   $(A + BC + G)$
   *(g) $IBM + K\bar{I}M + \bar{M}I + \bar{M}\bar{I}G + NI\bar{M}$
   *(h) $(I + C + B + \bar{M})(\bar{I} + M)(M + \dot{I} + L)(\bar{M} + \bar{I} + X)(V + I + M)$

5. Simplify:

(a) $AB + \bar{A}C\bar{D}E + \bar{B}C\bar{D}$

(b) $AB\bar{C}D + B\bar{C}\bar{E} + AE$

(c) $(A + B)(\bar{A} + C + \bar{D})(\bar{B} + C + \bar{D})$

(d) $(A + B + \bar{C})(B + \bar{C} + \bar{D})(A + D)$

(e) $\bar{A}B\bar{C} + AD + B\bar{C}\bar{D}$

(f) $\bar{A}B\bar{C} + \bar{A}BD + CD$

(g) $(\bar{A} + B + \bar{C})(C + D)(\bar{A} + B + \bar{D} + E)$

(h) $(\bar{A} + B + \bar{C})(\bar{A} + B + D + E)(C + D)$

6. Simplify:

(a) $AC + \bar{B}\bar{A} + \bar{D}\bar{C}B + \bar{C}EB + \bar{B}CF + B\bar{G}C$

(b) $(P + A + T)(P + E + \bar{T})(\bar{P} + O + T)(\bar{P} + U + \bar{T})(P + I)(\bar{I} + \bar{T})$

(c) $\bar{A}BC + CE + BCD + \bar{D}\bar{E}$

(d) $(\bar{A} + B)(C + A)(B + \bar{C} + D)(B + D + E)$

(e) $ABD + ACD + A\bar{B}E + AF + \bar{E}F$

(f) $\bar{K}L + \bar{L}M + HKM + \bar{G}\bar{M} + \bar{G}HJ$

(g) $\bar{X}ZY + \bar{Y}\bar{X}Z + \bar{Y}ZX + \bar{Z}YX + X\bar{Y}$

(h) $(\bar{A} + B)(A + \bar{C} + \bar{B})(B + A + \bar{C})(B + \bar{C} + \bar{A})(C + \bar{B} + \bar{A})$

*(i) $(B + A + D)(B + E + \bar{D})(\bar{B} + I + D)(\bar{B} + U + \bar{D})$
$$(\bar{B} + \bar{O})(O + D)$$

*(j) $IOU + \bar{U}E + AIO + \bar{E}O$

*(k) $ABC + DE + ACF + A\bar{D} + A\bar{B}\bar{E}$

*(l) $(\bar{X} + Y)(X + \bar{Z} + \bar{Y})(Y + X + \bar{Z})(Y + \bar{Z} + \bar{X})(Z + \bar{Y} + \bar{X})$

7. Transpose to a product of two expressions:

(a) $A(B + \bar{C}) + \bar{A}\bar{D}E$

(b) $(A + \bar{B}\bar{C})\bar{D} + DE(F + G)$

*(c) $(A + \bar{B})CD + \bar{C}(EF + \bar{G})$

*(d) $\bar{T}\bar{U}(\bar{V} + \bar{W}) + (XY + Z)U$

8. Transpose to a sum of two expressions:

(a) $(A + BC)(\bar{A} + \bar{D} + \bar{E})$

(b) $[A\bar{B} + \bar{C} + \bar{D}][D + (E + F)G]$

*(c) $[(A + \bar{B})C + D][F\bar{G} + \bar{D} + E]$

*(d) $[\bar{M}\bar{N} + \bar{O} + \bar{P}][(Q + R)S + O]$

9. Reduce the following to a single occurrence of A and $\bar{A}$. Express each as a product of two expressions and as a sum of two expressions.

(a) $AG + (A + B)C + \bar{A}D + (\bar{A} + F)E$

(b) $(A + \bar{B})(\bar{A} + C)(\bar{D} + E + AF)(G + \bar{H} + \bar{A}J)$

10. Find twelve ways to express with six literals, complementing variables only, i.e., without complementing products or sums:

$$\bar{A}C + \bar{A}B + A\bar{C} + A\bar{B}$$

Summary of Boolean Algebra Postulates and Theorems

Postulates

$X = 1$ or else $X = 0$

$1 \cdot 1 = 1$ $0 + 0 = 0$

$1 \cdot 0 = 0 \cdot 1 = 0$ $0 + 1 = 1 + 0 = 1$

$0 \cdot 0 = 0$ $1 + 1 = 1$

$\bar{1} = 0$ $\bar{0} = 1$

Theorems

1a. $0 \cdot X = 0$ **1b.** $1 + X = 1$

2a. $1 \cdot X = X$ **2b.** $0 + X = X$

3a. $XX = X$ **3b.** $X + X = X$

4a. $X\bar{X} = 0$ **4b.** $X + \bar{X} = 1$

5a. $XY = YX$ **5b.** $X + Y = Y + X$

6a. $XYZ = (XY)Z = X(YZ)$ **6b.** $X + Y + Z = (X + Y) + Z$
                                             $= X + (Y + Z)$

7a. $\overline{XY \ldots Z} = \bar{X} + \bar{Y} + \ldots + \bar{Z}$

7b. $\overline{X + Y + \ldots + Z} = \bar{X}\bar{Y} \ldots \bar{Z}$

8. $\bar{f}(X, Y, \ldots, Z, \cdot, +) = f(\bar{X}, \bar{Y}, \ldots, \bar{Z}, +, \cdot)$

9a. $XY + XZ = X(Y + Z)$ **9b.** $(X + Y)(X + Z) = X + YZ$

10a. $XY + X\bar{Y} = X$ **10b.** $(X + Y)(X + \bar{Y}) = X$

11a. $X + XY = X$ **11b.** $X(X + Y) = X$

12a. $X + \bar{X}Y = X + Y$ **12b.** $X(\bar{X} + Y) = XY$

12a′. $ZX + Z\bar{X}Y = ZX + ZY$

12b′. $(Z + X)(Z + \bar{X} + Y) = (Z + X)(Z + Y)$

13a. $XY + \bar{X}Z + YZ = XY + \bar{X}Z$

13b. $(X + Y)(\bar{X} + Z)(Y + Z) = (X + Y)(\bar{X} + Z)$

14a. $XY + \bar{X}Z = (X + Z)(\bar{X} + Y)$

14b. $(X + Y)(\bar{X} + Z) = XZ + \bar{X}Y$

15a. $X \cdot f(X, \bar{X}, Y, \ldots, Z) = X \cdot f(1, 0, Y, \ldots, Z)$

15b. $X + f(X, \bar{X}, Y, \ldots, Z) = X + f(0, 1, Y, \ldots, Z)$

16a. $f(X, \bar{X}, Y, \ldots, Z) = X \cdot f(1, 0, Y, \ldots, Z) + \bar{X} \cdot f(0, 1, Y, \ldots, Z)$

16b. $f(X, \bar{X}, Y, \ldots, Z) = [X + f(0, 1, Y, \ldots, Z)][\bar{X} + f(1, 0, Y, \ldots, Z)]$

2

Special Forms
of Boolean Expressions

Four forms of Boolean expressions that are of particular interest are

> Expanded sum of products
> Expanded product of sums
> "Minimum" sum of products
> "Minimum" product of sums

The expanded sum of products and expanded product of sums forms are useful for the analysis of Boolean expressions and their associated circuits and, also, they are a starting point for other methods of simplification which will be taken up later. The "minimum" sum of products and "minimum" product of sums forms are of interest because circuits are most frequently implemented directly from these expressions.

Expanded Sum of Products

In the expanded sum of products, each term contains every variable, either uncomplemented or complemented. To obtain the expanded sum of products from a sum of products, the missing variables are supplied in all possible combinations to each product. Actually, in so doing, Theorem

10a is used in reverse:

$$X = XY + X\bar{Y}$$

As an example, the sum of products

$$\bar{A}C\bar{D} + A\bar{B}D + A\bar{C}$$

will be expanded. The first term, $\bar{A}C\bar{D}$, has one missing variable B which is supplied in both its uncomplemented and complemented form; $\bar{A}C\bar{D}$ thus expands into two terms: $\bar{A}\bar{B}C\bar{D}$ and $\bar{A}BC\bar{D}$. The term $A\bar{B}D$ also expands into two terms: $A\bar{B}\bar{C}D$ and $A\bar{B}CD$. The $A\bar{C}$ term has two missing variables B and D. Two variables can occur in four possible combinations. Therefore, the term $A\bar{C}$ expands into four terms: $A\bar{B}\bar{C}\bar{D}$, $A\bar{B}\bar{C}D$, $AB\bar{C}\bar{D}$, and $AB\bar{C}D$. The $A\bar{B}\bar{C}D$ term has already been obtained by the expansion of the $A\bar{B}D$ term and is not repeated. The expanded sum of products is therefore

$$\bar{A}\bar{B}C\bar{D} + \bar{A}BC\bar{D} + A\bar{B}\bar{C}D + A\bar{B}CD + A\bar{B}\bar{C}\bar{D} + AB\bar{C}\bar{D} + AB\bar{C}D \quad (1)$$

Note, in this example, that the expanded sum of products contains seven of the sixteen possible combinations of the four variables. Although the "+" stands for the "inclusive or," the nature of an expanded sum of products is such that all terms are mutually exclusive; that is, if one of the terms equals 1, all others must equal 0. For instance, if $A = 0$, $B = 0$, $C = 1$, and $D = 0$, the first expanded product, $\bar{A}\bar{B}C\bar{D}$, is the only one equalling 1; all other expanded products will have one or more variables equalling 0, and thus all other products will equal 0. Of course it is possible, in an expanded sum of products, for all terms to equal 0.

Expanded Product of Sums

The expanded product of sums can be obtained from a product of sums in a similar manner. For example, the product of sums

$$(\bar{A} + \bar{B} + \bar{C})(\bar{A} + \bar{C} + D)(A + C)(A + \bar{D})$$

expands into

$$(\bar{A} + \bar{B} + \bar{C} + \bar{D})(\bar{A} + \bar{B} + \bar{C} + D)(\bar{A} + B + \bar{C} + D)(A + \bar{B} + C + \bar{D})$$
$$(A + \bar{B} + C + D)(A + B + C + \bar{D})(A + B + C + D)(A + \bar{B} + \bar{C} + \bar{D})$$
$$(A + B + \bar{C} + \bar{D}) \quad (2)$$

Here again, the missing variables are supplied in all possible combinations, this time to each sum.

The product of sums

$$(\bar{A} + \bar{B} + \bar{C})(\bar{A} + \bar{C} + D)(A + C)(A + \bar{D})$$

above, is equivalent to the sum of products

$$\bar{A}C\bar{D} + A\bar{B}D + A\bar{C}$$

used in the previous example. Equivalent expressions were purposely chosen for these examples to illustrate an important complementary relationship which will now be explained.

The expanded sum of products (1) contained seven of the sixteen possible combinations of four variables. A sum of the *other nine* combinations

$$\bar{A}\bar{B}\bar{C}\bar{D} + \bar{A}\bar{B}\bar{C}D + \bar{A}\bar{B}CD + \bar{A}B\bar{C}\bar{D} + \bar{A}B\bar{C}D$$
$$+ \bar{A}BCD + A\bar{B}C\bar{D} + ABC\bar{D} + ABCD \qquad (3)$$

represents the *complementary* expanded sum of products.

By definition, two expressions are complementary if, whenever one expression equals 1, the other equals 0 and vice versa, both expressions never both equalling 1 or both equalling 0. Of the sixteen possible combinations of four variables, one and only one combination will equal 1 at a given time, the other fifteen combinations equalling 0. Since all of the combinations are included in the two expanded sum of products (1) and (3), and no combination is included in both, one of the sums must equal 1 and the other must equal 0 at all times. Thus, the two expanded sums of products are complementary.

For example, if $\bar{A} = 1$, $\bar{B} = 1$, $C = 1$, and $\bar{D} = 1$, the first term in the original expanded sum of products equals 1, and the other six terms equal 0. Also, the nine terms in the complementary expanded sum of products equal 0. Thus, in this case the original sum equals 1 and the complementary sum equals 0.

If the complementary expanded sum of products (3) is complemented using DeMorgan's theorem, the expanded product of sums (2), equivalent to the original expanded sum of products (1), is obtained. Each sum in the expanded product of sums (2) is the complement of a combination (product) missing from the original expanded sum of products (1).

The expanded product of sums can therefore be obtained from the expanded sum of products by the complementation of the sum of all the missing products. Also, the expanded sum of products can be obtained from the expanded product of sums by the complementation of the product of all the missing sums.

Note that the *number of products* in the expanded sum of products *plus* the *number of sums* in the expanded product of sums *equals* 2^n, the total number of combinations of n variables. In the preceding example, for instance, there were seven products in the expanded sum of products and nine sums in the expanded product of sums; seven plus nine equals sixteen, the total number of combinations of four variables.

Each product in the expanded sum of products is also referred to as a "minterm." Each sum in the expanded product of sums is referred to as a "maxterm."

"Minimum" Boolean Expressions

The ultimate aim in a Boolean expression is a representation that corresponds to a minimum cost switching circuit. To this end, the minimum number of logic blocks is generally the primary criterion, and the minimum number of logic block inputs is typically the secondary criterion. (Other criteria may also be used, depending on the type of logic circuitry being used.)

As will be seen in the next chapter, logic blocks correspond to AND's and OR's in a Boolean sum of products or product of sums, the total number of which can be expressed as

1 + the number of terms of two or more literals[1]

The number of logic block inputs corresponds to

the number of literals in terms of two or more literals

+ the number of terms[1]

Broadly speaking, a minimum-term solution is therefore usually the primary objective. If more than one such solution can be realized, a minimum number of literals is usually the secondary objective. (A minimum-term solution may not be a minimum-literal solution, and vice versa.)

A minimum sum of products can be obtained from a sum of products by the application of the simplification theorems. A minimum product of sums can be similarly obtained from a product of sums. In practice, if the expression is quite complex it may not be easy to obtain a minimum by algebraic manipulation, or, even if a minimum is obtained, one may not be sure that it is a minimum. However, other methods of simplification, which will be taken up in Chapters 4 and 5, lead more systematically to a minimum.

A sum of products can be obtained from an expression not already in this form by simply "multiplying out" the expression, that is, applying Theorem 9a in reverse.

---

[1] Except for the special case of a single term.

A product of sums can be similarly obtained by the dual operation of "adding out," that is, applying Theorem 9b in reverse. However, since "multiplying out" is a more familiar operation than "adding out" (it being permissible in ordinary algebra), the following method for obtaining a product of sums form may be preferred:

1. Obtain a sum of products form by multiplying out.
2. Complement this expression, using DeMorgan's theorem. This complement will be in a product of sums form.
3. Multiply out this complement. The complement will now be in a sum of products form.
4. Complement this complemented sum of products form, using DeMorgan's theorem. Since the complement of a complement is equivalent to the original, a product of sums form equivalent to the original expression is obtained.

The procedure may be modified by utilizing the *dual* rather than the complement. Starting with a sum of products form, obtain the dual expression. The dual will be in a product of sums form. Multiply out the dual to get it in a sum of products form. Finally, obtain the dual of the dual to get a product of sums equivalent to the original. By using the dual, rather than the complement, it is not necessary to complement all literals twice in the procedure.

For an example, a product of sums will be obtained from the sum of products,

$$\bar{A}C\bar{D} + A\bar{B}D + A\bar{C}$$

Obtain the dual:

$$(\bar{A} + C + \bar{D})(A + \bar{B} + D)(A + \bar{C})$$

Multiply out the dual (some obvious simplifications can be made in the process):

$$\bar{A}\bar{B}\bar{C} + \bar{A}\bar{C}D + AC + A\bar{D} + \bar{B}\bar{C}\bar{D}$$

Obtain the dual of the dual to get the product of sums:

$$(\bar{A} + \bar{B} + \bar{C})(\bar{A} + \bar{C} + D)(A + C)(A + \bar{D})(\bar{B} + \bar{C} + \bar{D})$$

If a minimum product of sums is desired, the above expression is examined for further simplification, and it is found that either the first or last term is redundant. It is suggested that the reader verify this for practice.

There are thus two minimum product of sums:

$$(\bar{A} + \bar{B} + \bar{C})(\bar{A} + \bar{C} + D)(A + C)(A + \bar{D})$$
$$(\bar{A} + \bar{C} + D)(A + C)(A + \bar{D})(\bar{B} + \bar{C} + \bar{D})$$

Application of the simplification theorems can, of course, be made while the expression is still in the dual sum of products form.

By the judicious selection of pairs of terms to multiply together, the amount of work involved in the multiplying out process can be considerably reduced. In general, it is desirable to multiply together terms with variables in common, complemented or not.

For example, in the expression

$$(A + B + C)(D + E)(A + B + F)(\bar{D} + G)$$

multiplying together the first two terms, and multiplying together the last two terms, gives as a first step

$$(AD + AE + BD + BE + CD + CE)(A\bar{D} + AG + B\bar{D} + BG + \bar{D}F + FG)$$

whereas multiplying together the first and third terms (variables A and B common), and multiplying together the second and fourth terms (variable D common), gives as first step

$$(A + B + CF)(DG + \bar{D}E)$$

Earlier in the chapter, an expanded product of sums (2) was obtained from the first of the two minimum product of sums above. It is suggested that for practice the reader expand the second minimum product of sums and verify that the same expanded product of sums is obtained.

Minimum Factored Form

A minimum sum of products or a minimum product of sums can generally be factored. The factored form represents additional levels of logic blocks, with resulting longer circuit delays.

The minimum factored form is that expression with the absolute minimum number of literals. There is no formal method for always obtaining the minimum factored form. A minimum sum of products or minimum product of sums can be factored in all possible ways and the solution with the minimum number of literals selected. However, it may be possible to add redundancy before factoring and obtain an expression with fewer literals.

EXAMPLE

If the expression

$$VW + VX + W\bar{Y} + WZ + XYZ$$

is factored in all possible ways, the expression with the minimum number of literals is found to be

$$X(V + YZ) + W(V + \bar{Y} + Z) \quad \text{(8 literals)}$$

However, if a redundant Y is added to the WZ term, giving

$$VW + VX + W\bar{Y} + WYZ + XYZ$$

the minimum factored form

$$(X + W)(V + YZ) + W\bar{Y} \quad \text{(7 literals)}$$

is obtained.

Functions of *n* Variables

With n variables there are 2^n possible combinations, and these combinations can form $2^{(2^n)}$ different functions. For example, with two variables X and Y, there are four possible combinations:

$$\bar{X}\bar{Y}$$
$$\bar{X}Y$$
$$X\bar{Y}$$
$$XY$$

and these combinations can form sixteen different functions. These functions are shown, arranged in two columns so that each row contains a complementary pair.

Sixteen Functions of Two Variables

		0	$\bar{X}\bar{Y} + \bar{X}Y + X\bar{Y} + XY = 1$
		$XY = XY$	$\bar{X}\bar{Y} + \bar{X}Y + X\bar{Y} \qquad = \bar{X} + \bar{Y}$
	$X\bar{Y}$	$= X\bar{Y}$	$\bar{X}\bar{Y} + \bar{X}Y \qquad + XY = \bar{X} + Y$
$\bar{X}Y$		$= \bar{X}Y$	$\bar{X}\bar{Y} \qquad + X\bar{Y} + XY = X + \bar{Y}$
$\bar{X}\bar{Y}$		$= \bar{X}\bar{Y}$	$\qquad \bar{X}Y + X\bar{Y} + XY = X + Y$
		$X\bar{Y} + XY = X$	$\bar{X}\bar{Y} + \bar{X}Y \qquad = \bar{X}$
$\bar{X}Y$		$+ XY = Y$	$\bar{X}\bar{Y} \qquad + X\bar{Y} \qquad = \bar{Y}$
$\bar{X}\bar{Y}$		$+ XY = \bar{X}\bar{Y} + XY$	$\bar{X}Y + X\bar{Y} \qquad = \bar{X}Y + X\bar{Y}$

The table below gives a few corresponding values of n, 2^n, and $2^{(2^n)}$. Note that if the number of variables is increased by one, the number of functions is squared.

Number of Variables, n	Number of Combinations, 2^n	Number of Functions, $2^{(2^n)}$
0	1	2
1	2	4
2	4	16
3	8	256
4	16	65,536
5	32	4,294,967,296

PROBLEMS

1. Minimum sum of products: $\bar{A} + B\bar{C}$. Express as:
(a) minimum product of sums (making use of dual).
(b) expanded sum of products.
(c) expanded product of sums.

2. Express $A(B + \bar{C}) + D$ as:
(a) minimum sum of products.
(b) minimum product of sums.
(c) expanded sum of products.
(d) expanded product of sums.

*3. Minimum sum of products: $A\bar{C} + \bar{A}D + B\bar{D}$. Express as:
(a) minimum product of sums.
(b) expanded sum of products.
(c) expanded product of sums.

*4. Express $\bar{B}C + \bar{B}D + \bar{A}C + A\bar{D}$ as:
(a) minimum sum of products.
(b) minimum product of sums.
(c) expanded sum of products.
(d) expanded product of sums.

3

Logic Circuits and Diagrams

Before studying other methods of minimization, it will be well at this point (having related Boolean expressions to logic statements) to examine the relationships between Boolean expressions and logic circuits. Our primary goal, after all, is the minimization of a logic circuit, which is accomplished through the minimization of a related Boolean expression.

Also to be discussed in this chapter are the fundamental concepts of logic diagramming, a subject which, judging from the literature, is somewhat cloudy.

A logic circuit is represented by a logic diagram. We differentiate between two types of logic diagrams: the basic logic diagram and the detailed logic diagram.

Basic Logic Diagram

The basic logic diagram is of primary interest to the logic designer. It is here that he translates from the Boolean expression to a symbolic representation. It is a diagram of logic, not of any actual circuits, each function performed—AND, OR, and NOT—being represented by a logic symbol. Thus, the basic logic diagram depicts the logic functions in as simple a way as possible, with no reference to physical implementation.

The basic logic diagram symbols that will be used are shown in Fig. 3-1. The functions of these "logic blocks" are described. Conventionally, the logic flow, inputs to output, is from left to right.

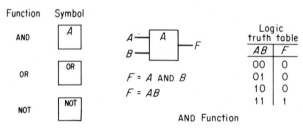

Figure 3-1

AND Function

Figure 3-2

AND Function

The AND output is 1 only if all of the inputs are 1. Conversely, the output is 0 only if one or more of the inputs are 0 (Fig. 3-2).

Where two-input functions are used for illustration, it should be recognized that the function also applies to more than two inputs.

OR Function

The OR output is 1 only if one or more of the inputs are 1. Conversely, the output is 0 only if all of the inputs are 0 (Fig. 3-3).

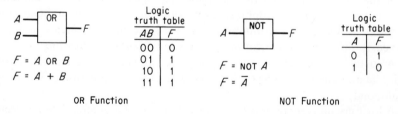

OR Function

Figure 3-3

NOT Function

Figure 3-4

NOT Function

The NOT output is 1 only if the input is 0. Conversely, the output is 0 only if the input is 1 (Fig. 3-4).

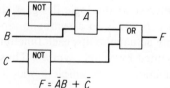

$F = \bar{A}B + \bar{C}$

Figure 3-5

Switching circuits are made up of interconnections of these logic blocks. A few examples of basic logic diagrams will now be given. The diagram for the expression $\bar{A}B + \bar{C}$ is shown in Fig. 3-5.

The diagram for the "exclusive or" function $A\bar{B} + \bar{A}B$ is shown in Fig. 3-6. A more

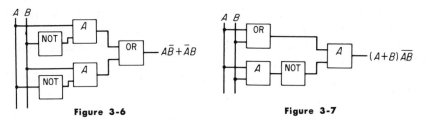

Figure 3-6   **Figure 3-7**

economical realization can be obtained (Fig. 3-7) by recognizing the follow-ing Boolean identities:

$$A\bar{B} + \bar{A}B = (A + B)(\bar{A} + \bar{B})$$
$$= (A + B)\overline{AB}$$

Detailed Logic Diagram

Since we are primarily interested in logic design, the basic logic diagram will be the one used in the following chapters of this book. However, the logic designer should have an understanding and appreciation of the physical implementation and how it is represented. The detailed logic diagram is of primary interest to the manufacturing and field-servicing personnel, where it is used, for example, for testing, maintenance, and troubleshooting diagnosis. The detailed logic diagram depicts both logic and nonlogic func-tions and includes the physical and electrical aspects of the circuit. The logic designer should know both how to produce a detailed logic diagram from a basic logic diagram and how to read a detailed logic diagram.

The logic functions are implemented by logic circuits or devices. The electrical condition information has been kept separate from the logic func-tions until this point; now, the two must be reconciled. Logic circuit inputs and outputs may be in either of two different states. These two states, some-times referred to as "on" and "off," may be typically two different voltage or current levels. One of the principal objectives at this point is to properly relate the logic 1 and 0 to, for example, the higher and lower voltage level of each logic circuit input and output. This relationship, which we shall refer to as *polarity*, is one of the most important aspects of the detailed logic diagram.

The actual voltage levels are immaterial; we represent the higher (more positive) of the two levels by **H**, and the lower (more negative) level by **L**. There are then two possible voltage/logic relationships:

H represents 1   or   **L** represents 1
L represents 0   **H** represents 0

If the more positive voltage is consistently selected as the 1-state, i.e., $\mathbf{H} = 1, \mathbf{L} = 0$, a circuit or system is said to have, or operate with, *positive logic.* If the more negative voltage is consistently selected as the 1-state, i.e., **L** = 1, **H** = 0, a circuit or system is said to have *negative logic.* If the voltage level/logic state assignment is not consistent, the circuit or system is said to have *mixed logic.*

Let us now look at some of the logic devices that are available for implementation.

AND Circuit

This device has the characteristic that the output is **H** only if all of the inputs are **H**. Conversely, the output is **L** only if one or more of the inputs are **L**. The voltage truth table for this device (with two inputs) is shown in Fig. 3-8(a).

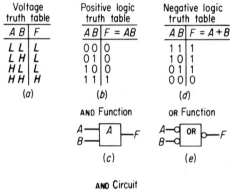

Voltage truth table			Positive logic truth table			Negative logic truth table		
A	B	F	A	B	F = AB	A	B	F = A + B
L	L	L	0	0	0	1	1	1
L	H	L	0	1	0	1	0	1
H	L	L	1	0	0	0	1	1
H	H	H	1	1	1	0	0	0
(a)			(b)			(d)		

AND Function   OR Function

(c)   (e)

AND Circuit

Figure 3-8

With positive logic, **H** = 1, **L** = 0, the logic truth table in Fig. 3-8(b) results, and the device is seen to perform the AND function. The output is **H** only if input *A* is **H** *and* input *B* is **H**. In the detailed logic diagram, the AND function implemented by an AND circuit is symbolized as shown in Fig. 3-8(c).

Logic circuits are customarily named according to the function they perform with positive logic—thus, the name "AND circuit" for this device. Logic devices, however, may perform more than one logic function. The AND circuit, with negative logic, **L** = 1, **H** = 0, performs the OR function, as seen by the logic truth table in Fig. 3-8(d). The output is **L** only if input *A* is **L** *or* input *B* is **L**.

In the detailed logic diagram, the assignment **L** = 1, **H** = 0 at any logic block input or output is designated by a small circle placed at the point where the input or output signal line joins the logic symbol. The absence of the circle indicates the assignment **H** = 1, **L** = 0 at a logic block input or output.

Thus, the OR function implemented by an AND circuit with negative logic would be symbolized as shown in Fig. 3-8(e).

OR *Circuit*

This device has the characteristic that the output is **H** only if one or more of the inputs are **H**. Conversely, the output is **L** only if all of the inputs are **L**. The voltage truth table for this device (with two inputs) is shown in Fig. 3-9, along with the logic truth tables with positive and negative logic and the associated detailed logic diagram symbols.

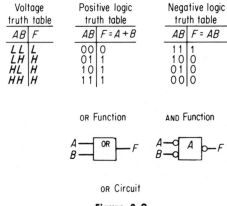

Voltage truth table		Positive logic truth table		Negative logic truth table	
AB	*F*	*AB*	*F = A + B*	*AB*	*F = AB*
LL	L	00	0	11	1
LH	H	01	1	10	0
HL	H	10	1	01	0
HH	H	11	1	00	0

OR Function   AND Function

OR Circuit

Figure 3-9

NAND *and* NOR *Functions*

The NAND function is described as follows. The NAND output is 0 only if all of the inputs are 1. Conversely, the output is 1 only if one or more of the inputs are 0 (Fig. 3-10). "NAND" is a contraction of "NOT AND". The NAND function is equivalent to a complemented AND, or NOT AND, function, e.g., NOT (*A* AND *B*). The NAND function is also called the *Sheffer Stroke* function, which has the Boolean symbol |. For example, $A \mid B$ is equivalent to $\overline{AB}$.

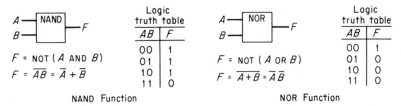

Logic truth table	
AB	*F*
00	1
01	1
10	1
11	0

$F = $ NOT (A AND B)

$F = \overline{AB} = \overline{A} + \overline{B}$

NAND Function

Figure 3-10

Logic truth table	
AB	*F*
00	1
01	0
10	0
11	0

$F = $ NOT (A OR B)

$F = \overline{A+B} = \overline{A}\,\overline{B}$

NOR Function

Figure 3-11

The NOR function is described as follows. The NOR output is 0 only if one or more of the inputs are 1. Conversely, the output is 1 only if all of the inputs are 0 (Fig. 3-11). "NOR" is a contraction of "NOT OR." The NOR function is equivalent to a complemented OR, or NOT OR, function, e.g., NOT (*A* OR *B*). The NOR function is also called the *Pierce Arrow* function, which has the Boolean symbol $\downarrow$. For example, $A \downarrow B$ is equivalent to $\overline{A + B}$.

Any Boolean function can be realized with NAND functions only or with NOR functions only.[1]

NAND Circuit

This device has the characteristic that the output is **L** only if all of the inputs are **H**. Conversely, the output is **H** only if one or more of the inputs are **L**. With positive logic, this device implements the NAND function; with negative logic, it implements the NOR function. However, it is more commonly used with mixed logic to implement the AND and OR functions, as shown in Fig. 3-12.

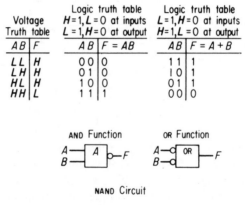

Voltage Truth table		Logic truth table $H=1, L=0$ at inputs $L=1, H=0$ at output		Logic truth table $L=1, H=0$ at inputs $H=1, L=0$ at output	
AB	*F*	*AB*	*F = AB*	*AB*	*F = A + B*
LL	*H*	00	0	11	1
LH	*H*	01	0	10	1
HL	*H*	10	0	01	1
HH	*L*	11	1	00	0

AND Function   OR Function

NAND Circuit

Figure 3-12

NOR Circuit

This device has the characteristic that the output is **L** only if one or more of the inputs are **H**. Conversely, the output is **H** only if all of the inputs are **L**. With positive logic, this device implements the NOR function; with negative logic, it implements the NAND function. However, it is more commonly used with mixed logic to implement the OR and AND functions, as shown in Fig. 3-13.

---

[1] Any Boolean function can also be realized with AND and NOT functions only, or with OR and NOT functions only.

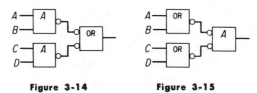

Voltage truth table	Logic truth table H = 1, L = 0 at inputs L = 1, H = 0 at output	Logic truth table L = 1, H = 0, at inputs H = 1, L = 0 at output
A B | F	AB | F = A + B	AB | F = AB
LL | H	00 | 0	11 | 1
LH | L	01 | 1	10 | 0
HL | L	10 | 1	01 | 0
HH | L	11 | 1	00 | 0

OR Function   AND Function

NOR Circuit

Figure 3-13

Note that a two-level NAND circuit is equivalent to an AND-OR circuit (Fig. 3-14) and that a two-level NOR circuit is equivalent to an OR-AND circuit (Fig. 3-15).

Figure 3-14   **Figure 3-15**

Note also that an entire symbol—the function being performed, in conjunction with the polarity assignment symbolized at the inputs and output—indicates the type of logic device. For example, the symbol in Fig. 3-16 represents a NOR circuit.

Figure 3-16

An important characteristic of the detailed logic diagram should be emphasized. The logic designer can look at the symbols and immediately recognize the logic functions being performed, without concern, if so desired, for what kinds of devices or what polarity signals are being used to implement the functions. On the other hand, the troubleshooter can look at the symbols and, observing the logic functions being performed and the signal polarities, can determine not only what kinds of devices are being employed but also the voltage levels to be expected, under given input conditions, at any point in the logic network.

For example, the symbol in Fig. 3-16 tells the logic designer that the AND function is being performed. Additionally, the small circles or absence of circles at the signal lines tell the troubleshooter that if both inputs are **L**, the output should be **H**. With this polarity information, the actual physical circuit can be tested to determine if it is functioning electrically the way the symbol says it should.

Inverter Circuit

This device has one input and one output, with the characteristic that the output is **L** only if the input is **H** and that the output is **H** only if the input is **L**. Either symbol in Fig. 3-17 is used for the inverter.

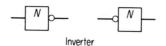

Inverter

Figure 3-17

We shall now discuss what is perhaps the most difficult detailed logic diagramming principle for designers to understand; it involves both the concepts of logic and their implementation with physical devices. More specifically, it involves the relationship between the *logic* NOT *function*, also referred to as *logic complementation* or *logic negation*, and the *electrical inverter circuit*, which performs *electrical inversion*.

As previously implied, the small circle at the input or output signal line of the inverter tells the user of the detailed logic diagram that when the input is **H**, the output should be **L**, and vice versa. Following the convention that the circle also indicates the **L** = 1, **H** = 0 polarity assignment, we find (see Fig. 3-18) that *no* logic transformation takes place; that is, the value of the logic output is identical to the value of the logic input. This is referred to as the *identity function* or *logic identity*. For example, if the function *A* is represented by **H** at the inverter input, the identical function *A* is represented by **L** at the inverter output.

The inverter circuit thus implements the identity function with mixed logic. (Physically, it is commonly a NAND or NOR circuit with only one input used.)

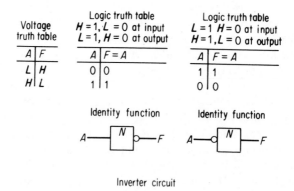

Inverter circuit

Figure 3-18

The reader should, at this point, dwell on the fact that the inverter is a nonlogic device and that electrical inversion, as symbolized by Fig. 3-19 in the detailed logic diagram, is *not* equated to logic negation, as symbolized by Fig. 3-20 in the basic logic diagram. The inverter changes the signal value but not the logic value. Logic negation changes the logic value but not the signal value. How do we reconcile the two?

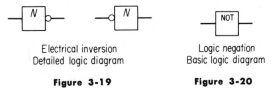

Electrical inversion
Detailed logic diagram

Figure 3-19

Logic negation
Basic logic diagram

Figure 3-20

Logic negation is not implemented by a circuit; it is implemented by a circuit; it is implemented by switching reference to the opposite polarity. For example, if the function A is represented by **H**, than A is represented by **L**. Logic negation is represented on the detailed logic diagram by a signal line with ends of opposite polarity, that is, with a small circle at one end and none at the other. This polarity difference may occur between two symbols or between an input or output and a symbol (See Fig. 3-21).

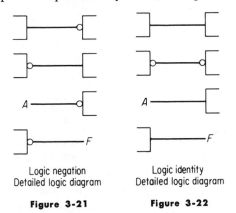

Logic negation
Detailed logic diagram

Figure 3-21

Logic identity
Detailed logic diagram

Figure 3-22

On the detailed logic diagram, a signal line with ends of like polarity represents logic identity (see Fig. 3-22).

In going from the basic logic diagram to the detailed logic diagram, how do we determine where inverters are needed? Start the detailed logic diagram by drawing the AND and OR symbols with the proper polarity inputs and outputs according to the logic circuits being used for implementation. Then, for each signal line, determine: (1) on the basic logic diagram, is there logic negation (the NOT function) or logic identity, and (2) on the detailed logic diagram, are the two ends of the line of opposite polarity (logic negation) or like polarity (logic identity), as indicated by the presence or absence of the small circles. If the two diagrams are not in agreement, an inverter is required in the signal line.

Basic Logic Diagram	*Detailed Logic Diagram*	*Inverter*
Logic identity	Logic identity (like polarity)	
Logic negation (NOT function)	Logic negation (opposite polarity)	
Logic identity	Logic negation (opposite polarity)	Required
Logic negation (NOT function)	Logic identity (like polarity)	Required

The insertion of an inverter in a signal line causes two effects: the inverter itself causes a change in the electrical signal value, and its insertion in the signal line causes a change in logic value. The inverter symbol is added in the detailed logic diagram, effecting a logic agreement between the two diagrams. Note that the inverter, in general, is used as often to change from a NOT function to an identity function as vice versa.

An example follows, in which the expression $F = \bar{A}B + \bar{C}$ is to be implemented. The basic logic diagram is shown in Fig. 3-23.

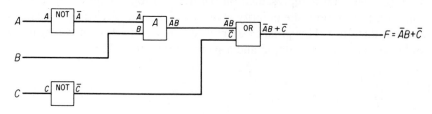

Basic logic diagram

Figure 3-23

Say it is desired to implement this function with NOR circuits. To start the detailed logic diagram, the AND and OR symbols are drawn with the proper polarity inputs and outputs according to NOR circuit implementation (Fig. 3-24). Now each signal line is examined. Between input A and the AND logic block there is logic negation, indicated by the NOT function in the basic logic diagram. Logic negation is also indicated on the detailed logic diagram, by opposite polarity: a small circle at the AND block and none at input A. This line, therefore, calls for no inverter.

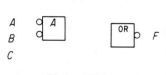

Figure 3-24

Between input B and the AND block there is logic identity (basic logic diagram); opposite polarity indicates logic negation on the detailed logic diagram. An inverter is therefore required in this line.

Between the AND block and OR block there is logic identity indicated by both diagrams; no inverter is required. Between input C and the OR block there is logic negation (basic logic diagram); like polarity indicates logic identity on the detailed logic diagram; an inverter is required. Last, between the OR block and output F there is logic identity (basic logic diagram); opposite polarity indicates logic negation on the detailed logic diagram; an inverter is required. The detailed logic diagram is shown in Fig. 3-25.

Note that even though a function may be implemented with NAND or NOR circuits, the logic designer does not have to "think" in terms of NAND's or NOR's; he can design in terms of AND's and OR's, with subsequent assignment of proper input and output polarities.

The detailed logic diagram can be read from either of two standpoints. First, the logic designer can trace the logic flow without regard to the devices or logic block polarities; he ignores all inverters (they are nonlogic devices), and any signal line with ends of opposite polarity represents logic negation.

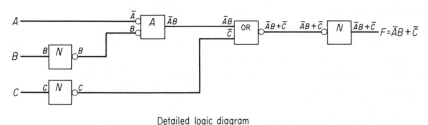

Detailed logic diagram

Figure 3-25

Let us trace the detailed logic diagram in Fig. 3-25 from a logic flow standpoint. The signal line between input A and the AND block has ends of opposite polarity, so we negate A and read "$\bar{A}$" at the upper AND block input. The signal line between B and the inverter has ends of like polarity, as does the line between the inverter and the AND block; we therefore read "B" at the lower AND block input. It now follows that we read "$\bar{A}B$" at the AND block output, and since the signal line between the AND block and the OR block has ends of like polarity, we read "$\bar{A}B$" at the upper OR block input.

The signal line between C and the inverter has ends of like polarity, so we read "C" at the inverter input. However, the signal line between that inverter and the OR block has ends of opposite polarity, so we read "$\bar{C}$" at the lower OR block input.

It now follows that we read "$\bar{A}B + \bar{C}$" at the OR block output, and since the signal line between the OR block and the following inverter has ends of like polarity, as does the signal line between that inverter and output F, we read "$\bar{A}B + \bar{C}$" for the output F.

A word now about the choice between the two inverter symbols. If an inverter is inserted in a signal line with ends of opposite polarity, it is desirable that the inverter symbol be chosen so that the two resulting signal lines (one on each side of the inverter) have ends of like polarity. For example, between input B and the AND block in Fig. 3-25, the inverter symbol was

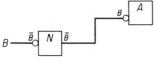

Figure 3-26

particularly chosen according to this criterion. The other symbol would have resulted in the configuration in Fig. 3-26, in which the signal line between input *B* and the inverter has ends of opposite polarity, as does the signal line between the inverter and the AND block.

Why is this choice important? Between input *B* and the AND block, for instance, if the inverter symbol in Fig. 3-26 had been chosen, the logic flow would have been as follows. The signal line between input *B* and the inverter has ends of opposite polarity, so we negate *B* and read "$\bar{B}$" at the inverter input. The signal line between the inverter and the AND block has ends of opposite polarity, so we negate $\bar{B}$ and read "*B*" at the AND block input. The improper choice of inverter symbol thus still "works," but it causes us to do a needless double negation.

The proper choice of inverter symbol was also important between the OR block and output *F* in Fig. 3-25.

If an inverter is inserted in a signal line with ends of like polarity, the choice of inverter symbol is arbitrary. Between input *C* and the OR block, in Fig. 3-25, this is the case.

The detailed logic diagram can also be read from a second standpoint. The troubleshooter can read the input and output polarity symbols and thus trace voltage levels. For example, in Fig. 3-25, if both inputs of the AND block are **L**, the AND block output should be **H**. If either or both of the inputs of the OR block are **H**, the output should be **L**. If the input of an inverter is **H**, the output should be **L**, and vice versa. If these polarities are traced through from inputs to output, it can be seen that for output *F* to be **H**, input *A* must be **L** and input *B* must be **H**, or input *C* must be **L**.

Standard Symbols

A few words about standard symbols. Two major standards exist: *American Standard Graphic Symbols for Logic Diagrams*, AIEE No. 91—ASA Y32.14, and *Military Standard Graphic Symbols for Logic Diagrams*, MIL-STD-806B. The two standards differ in many important respects and cannot be used interchangeably. Industry is not in full agreement on either of these standards, and many company standards exist which are different from both in varying degrees.

While these standards are of principal importance in detailed logic diagrams, some of their characteristics, and some similarities and differences between the two standards may be of interest to the reader.

Some of the standard symbols are shown in Fig. 3-27. Although input and output lines are shown here for instructional purposes, the symbols themselves do not include the lines.

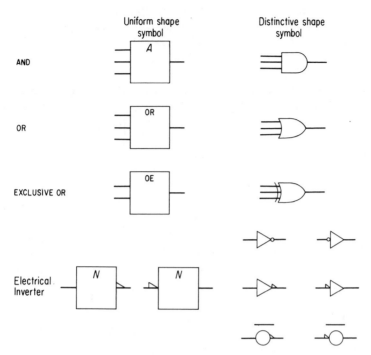

Figure 3-27

The AIEE-ASA standard allows both the uniform shape and the distinctive shape symbols; MIL-STD allows only the distinctive shape. The MIL-STD uses the small circles to indicate the $L = 1$, $H = 0$ polarity assignment; AIEE-ASA uses a small open right triangle. AIEE-ASA also uses a (redundant) small filled right triangle to indicate the $H = 1$, $L = 0$ polarity assignment. To further complicate the issue, AIEE-ASA uses the small circle to indicate logic negation, giving this standard another redundancy in a choice of two ways to indicate logic negation (signal lines with ends of opposite polarity being the other).

The choice of the standard uniform shape electrical inverter symbol is unfortunate, since the "N" connotates the logic NOT function, with which the inverter should not be confused. "I" for "inverter" or "identity" would have been better. In this book, the nonstandard basic logic diagram symbol for the NOT function (Fig. 3-28) was purposely chosen for differentiation with the inverted symbol.

Figure 3-28

The fundamental concepts underlying both standards, and features of both, were incorporated in this chapter.

It should be realized that in practical situations the designer may be concerned with other details not discussed here, such as physical limits on the number of allowable logic device inputs (fan in), the types and number of other devices that can be driven from a logic device output (fan out), socket locations, pin numbers, test points, and so forth.

As industry and technology continually advance at a rapid pace, the hardware from which circuits are built are ever changing. For this reason, and because logic rather than hardware is our primary interest, the internal structure of the devices discussed in this chapter are not covered. Much literature exists on the subject for the interested reader.

4

Tabular and Consensus Methods of Minimization

Optional Combinations

Until now, for a desired circuit function, all of the possible input combinations could be considered as being divided into two groups: one group consisting of those combinations for which a circuit output is desired, and the other group consisting of those combinations for which no output is desired.

For example, suppose a circuit output is specified by

$$\bar{A}\bar{C} + ABC$$

This expression expands into

$$\bar{A}\bar{B}\bar{C} + \bar{A}B\bar{C} + ABC$$

No output is desired for the remaining five possible combinations:

$$\bar{A}\bar{B}C + \bar{A}BC + A\bar{B}\bar{C} + A\bar{B}C + AB\bar{C}$$

which can simplify to

$$\bar{A}C + A\bar{C} + A\bar{B}$$

or

$$\bar{A}C + A\bar{C} + \bar{B}C$$

Note that the expressions for *output* and *no output* are complementary.

Now, all of the possible input combinations will be considered as being divided into *three* groups: one group consisting of those combinations for which an output is desired, another group of those combinations for which no output is desired, and a third group of *optional* combinations.

These optional input combinations can arise from two possible conditions:

1. The optional combinations are invalid; that is, they are known never to occur.
2. The optional combinations are "don't care" combinations; that is, we do not care whether or not we get an output if these input combinations occur.

Optional combinations, if any, arise as part of the circuit specifications. It is not necessary to differentiate between invalid and don't care combinations; they both influence a circuit expression in the same way. If optional combinations are added to a Boolean expression for a circuit output, the expression may become simpler or more complicated. It must therefore be determined which optional combinations to add, and which not to add, in order to achieve optimum simplification.

EXAMPLES

(a) There are two keys A and B. It is desired that these keys light a lamp only if key A is on and key B is not on. The Boolean expression for lighting the lamp is therefore $A\bar{B}$. The keys are mechanically interlocked so that it is impossible for both keys to be on together. This means that *the combination AB can never occur*. If this optional combination AB is added to the output expression $A\bar{B}$, the expression $A\bar{B} + AB = A$ results. This simplification can be seen intuitively: If key A is on, it is not necessary to stipulate that key B be not on; it cannot be on since keys A and B cannot be on together because of the interlock.

    In this example, utilization of the optional combination led to simplification. In the next example, it will lead to complication.

(b) Using the same two interlocked keys, suppose now that the lamp is to light only when neither key is on. The Boolean expression for lighting the lamp is $\bar{A}\bar{B}$. If the optional combination AB is added to the output expression $\bar{A}\bar{B}$, the expression $\bar{A}\bar{B} + AB$ results. Since utilization of the optional combination complicates (rather than simplifying) the expression, it is better not to add the optional combination, but leave the expression in its original form, $\bar{A}\bar{B}$.

In these two examples only one optional combination was involved, and it was not very much work to investigate whether or not its utilization led to simplification.

Now, an example with two optional combinations will be examined. The expression for the output is

$$AC\bar{D} + BC\bar{D}$$

and the combinations $\bar{A}\bar{B}\bar{C}D$ and $\bar{A}\bar{B}C\bar{D}$ are given as optional.

Using neither optional combination, the original expression is retained:

$$AC\bar{D} + BC\bar{D} \tag{1}$$

Using both optional combinations,

$$AC\bar{D} + BC\bar{D} + \bar{A}\bar{B}\bar{C}D + \bar{A}\bar{B}C\bar{D} = C\bar{D} + \bar{A}\bar{B}\bar{C}D \tag{2}$$

an expression of the same complexity is obtained.

Using just the optional combination $\bar{A}\bar{B}\bar{C}D$,

$$AC\bar{D} + BC\bar{D} + \bar{A}\bar{B}CD \tag{3}$$

a more complicated expression results.

Finally, using only the optional combination $\bar{A}\bar{B}C\bar{D}$,

$$AC\bar{D} + BC\bar{D} + \bar{A}\bar{B}C\bar{D} = C\bar{D} \tag{4}$$

maximum simplification is achieved.

With two optional combinations, four trials were necessary to determine the optimum solution. In general, with n optional combinations, 2^n such trials are necessary. Obviously, n does not have to be very large before the work involved in this sort of algebraic simplification becomes prohibitive.

The tabular and map methods of minimization, however, handle optional combinations with facility, as will be seen in this and the following chapters.

Tabular Method of Minimization

The tabular method of minimization is based principally on the theorem

$$XY + X\bar{Y} = X$$

X representing one or more variables, and Y representing a single variable.

The first step in the tabular method is to get the expression (to be simplified) in the expanded sum of products form.[1] The preceding theorem is then applied exhaustively to obtain all irreducible terms, that is, terms to which the theorem cannot be further applied.

The theorem is applied first to all possible pairs of terms. Two terms to which the theorem can be applied will reduce to one term which is smaller by one literal. For example,

$$ABC + AB\bar{C} = AB$$

Next, all terms reduced by one literal are examined to see whether they can be combined further, by the application of the theorem, to reduce to a still smaller term containing two fewer literals than the original terms. This procedure is continued until no further terms can be combined. The resulting irreducible terms are called "prime implicants."

The last step in the method is to select in an optimum manner prime implicants that account for all of the original expanded terms. These prime implicants will form a minimum sum of products.

In the reduction process, the following relationships hold:

$n =$ number of variables
$m =$ number of variables occurring in all possible combinations in 2^m terms
$(n - m) =$ number of variables constant in the 2^m terms
The 2^m terms reduce to a single term defined by the constant $(n - m)$ variables.

EXAMPLE

$$A\bar{B}C\bar{D}\bar{E} + A\bar{B}C\bar{D}E + A\bar{B}CD\bar{E} + A\bar{B}CDE = A\bar{B}C$$

n (the number of variables) $= 5$,
m (the number of variables occurring in all possible combinations) $= 2$ (D and E); these combinations occur in $2^m = 2^2 = 4$ terms.

The remaining $(n - m = 3)$ variables are constant in the four terms, and the expression reduces to $A\bar{B}C$.

In the tabular method of minimization, the expression to be simplified must first be expanded, if it is not already in expanded form. However, instead of the expanded expression being written in algebraic form, it can be written in tabular form. For example, the sum of minterms

$$\bar{A}\bar{B}CD + \bar{A}B\bar{C}D + \bar{A}BC\bar{D} + AB\bar{C}\bar{D}$$

---

[1]It will be seen later that the expanded product of sums form can also be used.

can be written as

$$\bar{A}\bar{B}CD$$

$$\bar{A}B\bar{C}D$$

$$\bar{A}BC\bar{D}$$

$$AB\bar{C}\bar{D}$$

This table can be written in a still more convenient form by simply using *A*, *B*, *C*, and *D* as columnar headings, and then, in the table, using a 1 to represent an uncomplemented literal and a 0 to represent a complemented literal. The preceding table would then be written as

A	B	C	D
0	0	1	1
0	1	0	1
0	1	1	0
1	1	0	0

In the explanation of the tabular method, the following expression will be used as an example:

$$AB + A\bar{B}C\bar{D} + A\bar{B}\bar{C}D + \bar{A}BCD + \bar{A}B\bar{C}\bar{D}$$

The first step is to expand this expression into a table. The *AB* term will expand into four terms; the last four terms are already in expanded form, and none of these is the same as a term resulting from the expansion of *AB*. Therefore, the table has eight rows, as shown below:

A	B	C	D
1	1	0	0
1	1	0	1
1	1	1	0
1	1	1	1
1	0	1	0
1	0	0	1
0	1	1	1
0	1	0	0

Instead of examining all possible pairs of rows for application of the theorem, the work can be simplified by the following reasoning. For two rows to combine, they must differ in only one column; in one row, that column must contain a 0; in the other row, that column must contain a 1. Thus, a necessary condition for two rows to combine is that one of the rows must contain one more 1 than the other row. If, therefore, the rows are grouped according to the number of 1's per row, and the groups are arranged consecutively according to the number of 1's per row, it is necessary only to compare rows in one group with rows in an adjacent group. The table is thus reordered. Lines are drawn between adjacent groups to aid in identification.

		A	B	C	D
One 1 per row	{	0	1	0	0
	⎧	1	1	0	0
Two 1's per row	⎨	1	0	1	0
	⎩	1	0	0	1
	⎧	1	1	0	1
Three 1's per row	⎨	1	1	1	0
	⎩	0	1	1	1
Four 1's per row	{	1	1	1	1

Now we look for rows that combine. The 0100 row is compared with each of the three rows in the next group. The 0100 and 1100 rows differ in only one column, column A; therefore, these two rows combine into a row which is written, in a second table, as —100.

A	B	C	D		A	B	C	D
0	1	0	0✓		—	1	0	0
1	1	0	0✓		1	1	0	—
1	0	1	0✓		1	1	—	0
1	0	0	1✓		1	—	1	0
					1	—	0	1
1	1	0	1✓					
1	1	1	0✓		1	1	—	1
0	1	1	1✓		1	1	1	—
					—	1	1	1
1	1	1	1✓					

The entry —100 records that the terms $\bar{A}B\bar{C}\bar{D}$ and $AB\bar{C}\bar{D}$ reduce to $B\bar{C}\bar{D}$. Since the 0100 and 1100 rows are accounted for by the new row —100, they are "checked off," signifying that they are *not* prime implicants.

Since *all* of the prime implicants must be found, we continue to look for all possible combinations of rows, even with rows that have already been checked off. The 0100 row and the 1010 row are compared and it is found that they differ in more than one column; therefore, they do not combine. The 0100 row and the 1001 row are compared, and it is found that they do not combine either. Lines are drawn between adjacent groups in all tables; therefore, a line is drawn under the —100 row in the second table.

The three rows in the second group are now compared with the three rows in the third group. Rows 1100 and 1101 combine to give 110—; rows 1100 and 1110 combine to give 11—0; 1010 and 1110 combine to give 1—10; and rows 1001 and 1101 combine to give 1—01. All other pairs of rows, one from the second group and one from the third group, differ in more than one column, and thus do not combine.

Next, the three rows in the third group are compared with the 1111 row in the last group. Note that all rows with a single 0 will combine with an all-1 row. Thus, new rows 11—1, 111—, and —111 are obtained. It should also be noted that a row with all 0's would combine with all rows containing a single 1.

In this example, all rows in the original table have combined and have been checked off. If any row had not combined, it would have been a prime implicant.

The next step is to compare the rows in the second table in search for further combinations. Again, for two rows to combine, they must differ in only one column; in one row, that column must contain a 0; in the other row, that column must contain a 1. All other columns must be identical; that is, in all other columns, both rows must contain 0's, both rows must contain 1's, or both rows must contain —'s. Note especially that a — in one row must match with a — in another row.

The —'s, in fact, speed up the comparison process. For instance, in the only row in the first group of the second table, —100, there is a — in the A column. In the next group, none of the four rows have a — in the A column. Therefore, it can be seen immediately that —100 does not combine and is a prime implicant. A prime implicant is identified by an asterisk, as shown below:

A	B	C	D			A	B	C	D	
—	1	0	0	*		1	1	—	—	*
1	1	0	—	√						
1	1	—	0	√						
1	—	1	0	*						
1	—	0	1	*						
1	1	—	1	√						
1	1	1	—	√						
—	1	1	1	*						

Next, the four rows in the second group are compared with the three rows in the third group: 110— combines with 111— to give 11— — in a third table; 11—0 combines with 11—1 to also give 11 — —. This 11— — row is not repeated, but it serves to check off two more rows in the table. 1—10 and 1—01 have —'s in the B column, and since there are no —'s in the B column in any row in the third group, it is immediately seen that these two rows are prime implicants. —111 is also a prime implicant. In the third table there is only a single row, which obviously cannot combine with any other row, and so it is also a prime implicant.

Now that all prime implicants have been obtained, the last step is to select in an optimum manner prime implicants that account for all of the original expanded terms. To do this, a different kind of table, called a prime implicant table, is constructed: There is a column for each of the original terms, and a row for each prime implicant. For each prime implicant, a check mark is placed in the columns of those terms accounted for by that prime implicant. The completed table is shown below.

1100	1101	1110	1111	1010	1001	0111	0100	
✓							✓	—100 *
		✓	✓					1—10 *
	✓				✓			1—01 *
			✓			✓		—111 *
✓	✓	✓	✓					11— —

A prime implicant with no —'s will account for only one term; a prime implicant with one — will account for two terms; a prime implicant with two —'s will account for four terms; etc. For instance, the first prime implicant —100 ($B\bar{C}\bar{D}$) accounts for two terms 1100 ($AB\bar{C}\bar{D}$) and 0100 ($\bar{A}B\bar{C}\bar{D}$).

Although there is a formal method for determining the optimum selection of prime implicants, a better understanding of the problem can be gained if it is first looked at intuitively. First, any columns with only a single check mark are noted. A single check mark indicates that there is only one prime implicant that will account for the term in that column; therefore, the prime implicant in that row is required in the final expression—it is an "essential prime implicant."

In the example, the last four columns have only a single check mark. The term 1010 is accounted for only by the prime implicant 1—10; 1001, by the prime implicant 1—01; 0111, by the prime implicant —111; and 0100, by the prime implicant —100. The first four prime implicants are therefore required. Required prime implicants are identified by an asterisk. The terms are checked off as they are accounted for. Prime implicant —100 accounts for the terms 1100 and 0100; 1—10, for the terms 1110 and 1010; 1—01, for 1101 and 1001; and —111, for 1111 and 0111. At this point, it can be seen that all terms have been accounted for, and the first four prime implicants are the only ones required.

Forming a sum of products with these four prime implicants gives the minimum sum of products equivalent to the original expression

$$B\bar{C}\bar{D} + AC\bar{D} + A\bar{C}D + BCD$$

This particular problem was easy because the accounting for the columns with single check marks effected the solution. In some problems, many or all columns may have many check marks, making the solution of this last step by intuitive methods more difficult. In the next example a formal method of accomplishing this last step will be examined, as well as the method for treating optional combinations.

Optional Combinations with Tabular Method

Any optional combinations are added at the bottom of the original table and, for identification purposes, a line is drawn separating them from the valid combinations. The optional combinations and valid combinations are not differentiated again until after all prime implicants have been obtained; that is, in reordering the table and finding all prime implicants the optional combinations and valid combinations are treated alike.

After the prime implicants have been obtained, the prime implicant table is constructed with columns for the valid combinations only, since only the valid combinations must be accounted for. The optional combinations are thus used only for the possible generation of additional prime implicants, or prime implicants with fewer literals.

EXAMPLE

In the following example there are nine valid combinations and two optional combinations, 0000 and 0100. It is suggested that, for practice, the reader carry out the steps shown without referring to the book, and then check his results. Note that in the prime implicant table there are columns only for the valid combinations. The optional combinations account for the two missing check marks in the third row and the one missing check mark in the fourth row.

Example with Optional Combinations

A B C D	A B C D	A B C D	A B C D
0 0 0 1	0 0 0 0 ✓	0 0 0 — ✓	0 — 0 — *
0 0 1 1	—	0 — 0 0 ✓	—
—	0 0 0 1 ✓	—	— 1 0 — *
0 1 0 1	0 1 0 0 ✓	0 0 — 1 *	—
1 0 1 0	—	0 — 0 1 ✓	1 — 1 — *
1 0 1 1	0 0 1 1 ✓	0 1 0 — ✓	1 1 — — *
1 1 0 0	0 1 0 1 ✓	— 1 0 0 ✓	
1 1 0 1	1 0 1 0 ✓	— 0 1 1 *	
1 1 1 0	1 1 0 0 ✓	— 1 0 1 ✓	
1 1 1 1	—	1 0 1 — ✓	
—	1 0 1 1 ✓	1 — 1 0 ✓	
0 0 0 0	1 1 0 1 ✓	1 1 0 — ✓	
0 1 0 0	1 1 1 0 ✓	1 1 — 0 ✓	
	—	—	
	1 1 1 1 ✓	1 — 1 1 ✓	
		1 1 — 1 ✓	
		1 1 1 — ✓	

	0001	0011	0101	1010	1011	1100	1101	1110	1111	
	✓	✓								0 0—1 U
		✓			✓					—0 1 1 V
	✓		✓							0—0— W
			✓			✓	✓			—1 0— X
				✓	✓			✓	✓	1—1— Y
						✓	✓	✓	✓	1 1—— Z

Algebraic Solution of Prime Implicant Table

The formal solution of the prime implicant table, interestingly enough, utilizes Boolean algebra. The prime implicants are the variables, and are given letter names; in the example, they are designated U, V, W, X, Y, and Z, respectively.

For each valid combination, a Boolean expression is written indicating which prime implicants can account for it; this expression, by its nature, will be a sum. Thus, in the example, the combination 0001 can be accounted for by the prime implicants U or W, which is written in Boolean algebra as $(U + W)$; the combination 0011 can be accounted for by the prime implicants U or V, that is, $(U + V)$; etc.

A product of all these sums is formed, the resulting Boolean expression indicating how all of the combinations in the table can be accounted for. In the example, the product of sums obtained is

$$(U + W)(U + V)(W + X)(Y)(V + Y)(X + Z)(X + Z)(Y + Z)(Y + Z)$$

Some obvious simplification gives

$$(U + W)(U + V)(W + X)(Y)(X + Z)$$

This expression "says" that all the combinations in the table can be accounted for by the prime implicants $(U$ or $W)$ and $(U$ or $V)$ and $(W$ or $X)$ and (Y) and $(X$ or $Z)$.

The product of sums is now multiplied out, making use of the simplification theorems whenever possible. Note that since there are no complemented variables involved, only a few of the simplification theorems need be considered. Also, one can be selective in which terms he chooses to multiply out first; if those with the most literals in common are multiplied together first, the process is simplified.

Multiplying out gives an *equivalent* expression in the sum of products form:

$$
\begin{aligned}
(U + W)(U + V)(W + X)(Y)(X + Z) \\
= (U + VW)(X + WZ)Y \\
= UXY + UWYZ + VWXY + VWYZ
\end{aligned}
$$

This sum of products expression logically states the same thing as the previous product of sums expression, except in another way: It states that all the combinations in the table can be accounted for by the prime implicants $(U$ and X and $Y)$ or $(U$ and W and Y and $Z)$ or $(V$ and W and X and $Y)$ or $(V$ and W and Y and $Z)$. In general, the smallest term is selected to account for the table, since it represents the fewest required prime implicants.

In the example, the UXY term is selected because it is the smallest, and the prime implicants

$$U = 00\text{—}1 \ = \bar{A}\bar{B}D$$

$$X = -10-- = B\bar{C}$$
$$Y = 1-1-- = AC$$

are used to account for the table. If any other term had been selected, four rather than three prime implicants would have been required.

The selected prime implicants are summed to obtain the minimum sum of products, and the solution is

$$\bar{A}\bar{B}D + B\bar{C} + AC$$

This method of solution of the table not only gives *one* minimum sum of products, it gives *all* possible minimums if there are more than one. Furthermore, it gives *all irredundant solutions*, that is, all solutions from which no prime implicant may be removed and still have all output combinations accounted for. In the preceding example, there are four irredundant solutions.

Simplification of Prime Implicant Table

Prior to the algebraic solution, the prime implicant table can often be simplified by the elimination of certain columns and rows, as follows.

1. For any essential prime implicant, the associated row can be eliminated, along with every column that the prime implicant accounts for. The essential prime implicant is later added to the expression obtained from the solution of the reduced table.

2. A column, *a*, can be eliminated if it has check marks in every row that some other column, *b*, has. (Column *b* represents a "tighter" requirement than column *a*; if column *b* is accounted for, column *a* will be also.)

EXAMPLES

(a)

	a	b	
	✓	✓	
	✓	✓	
	✓		

Column *a* can be eliminated because it has check marks in every row that column *b* has.

(b)

	a	b
	✓	✓
	✓	✓

Either column, a or b, can be eliminated because each has check marks in every row that the other has.

3. A row, z, can be eliminated if some other row, y, has check marks in every column that z has, *and* if the number of literals in the z prime implicant is equal to or greater than the number of literals in the y prime implicant. (The y prime implicant is "stronger" than the z prime implicant in that it accounts, at least, for all columns that z does, and at the same time does not require more literals than z.)

EXAMPLES

(a)

	✓	✓	$-$ 1 0 1 y
	✓		1 $-$ 0 1 z

Row z can be eliminated because row y has check marks in every column that row z has, and the number of literals in the z prime implicant is equal to or greater than the number of literals in the y prime implicant.

(b)

	✓	✓	$-$ 1 0 $-y$
	✓	✓	$--$ 0 1 z

Either row, y or z, can be eliminated because each has check marks in every column that the other has, and both prime implicants have the same number of literals.

(c)

✓	✓	— 1 — — y
✓	✓	— — 0 1 z

Only row z can be eliminated because even though each row has check marks in every column that the other has, the z prime implicant has a greater number of literals than the y prime implicant.

(d)

✓	✓	— 1 0 1 y
✓		1 1 — — z

Neither row can be eliminated. Even though row y has check marks in every column that row z has, the number of literals in the z prime implicant is less than the number of literals in the y prime implicant.

If the number of literals is of no consequence, a row, z, can be eliminated simply if some other row, y, has check marks in every column that z has, irrespective of the number of literals in the y and z prime implicants. In such a case, in the preceding example (d), row z could be eliminated.

In general, not all possible minimum and irredundant solutions will be obtained if row elimination (rule 3) is employed.

These table simplification rules should be applied in the order shown:

1. Elimination of rows and columns associated with essential prime implicants. These prime implicants must be added to the expression obtained from the solution of the reduced table.
2. Column elimination.
3. Row elimination.

Column elimination should be done before row elimination, since the elimination of a column can result in the elimination of a row, but not vice versa.

Row elimination may create further essential prime implicants. If so, the rules are applied again in the same order.

When no further table simplification is possible, the table may have been completely solved; if not, the algebraic method can be applied to the reduced table remaining.

As a final example, the simplification rules will be applied to the table on page 58. Column 1010 has only a single check mark, which is in the Y row;

prime implicant Y is thus essential. The Y prime implicant also accounts for the columns 1011, 1110, and 1111. Row Y and the four columns it accounts for are eliminated from the table. Prime implicant Y must appear in the solution. The reduced table, at this point, appears as

0001	0011	0101			1100	1101			
√	√								0 0 — 1 U
	√								— 0 1 1 V
√		√							0 — 0 — W
		√			√	√			— 1 0 — X
					√	√			1 1 — — Z

Either column 1100 or 1101 can be eliminated; arbitrarily, column 1101 will be eliminated.

Row U can cause row V to be eliminated. Row X can cause row Z to be eliminated.

The table at this point appears as

0001	0011	0101			1100				
√	√								0 0 — 1 U
√		√							0 — 0 — W
		√			√				— 1 0 — X

Column 0011 has only a single check mark, making prime implicant U essential. Column 1100 has only a single check mark, making prime implicant X essential. Prime implicants U and X together account for all of the remaining columns, and the table is completely solved. The solution UXY agrees with the minimum previousy obtained by the algebraic method.

Weighting of Prime Implicants

If there is more than one solution with a minimum number of prime implicants, the solution with the least number of literals is usually desired.

As an aid in obtaining the desired solution, each prime implicant can be assigned a "weight" according to the number of literals it contains, and the solution with the smallest weight selected.

EXAMPLE

Given the following prime implicants, the weights would be assigned as shown:

		Weight
U	1 1 1 — —	3
V	1 1 — 0 —	3
W	1 — 1 1 —	3
X	— 0 — 1 —	2
Y	— — 0 0 —	2
Z	— — — — 0	1

Given the following terms in the algebraic solution, each solution would have weights as shown:

$$VW + UVX + UWY + UXY + VXZ + WYZ + XYZ$$
Weight: 6   8   8   7   6   6   5

Note that it is possible that a solution with the minimum number of prime implicants may contain more literals than a solution containing more prime implicants. If a solution containing a minimum number of literals is desired, this method of weighting points out which solution to choose.

Complementary Approach with Tabular Method

In the complementary approach, a table is made up of those combinations for which *no output* is desired. Any optional combinations are added to the table as before. The table is solved in the usual manner, and the resultant minimum sum of products is complemented using DeMorgan's theorem. Thus, the final solution appears in a minimum product of sums form.

Since either the direct or the complementary approach may lead to a better solution, it is often desirable to try both solutions, selecting the optimum one. If the number of "output" combinations is large compared to the number of "no-output" combinations, the complementary approach may be used simply to reduce the labor involved in reaching a "good" solution.

If there are no optional combinations, the minimum product of sums obtained using the complementary approach is a true equivalent of the minimum sum of products obtained with the direct approach. However, if there are optional combinations involved, the two expressions *may* not be truly equivalent.

The two expressions will always be equivalent in the sense that if any "output" combination equals 1, both expressions will equal 1, and if any "no-output" combination equals 1, both expressions will equal 0. However, if an optional combination equals 1, the two expressions may or may not be equivalent.

If, with optional combinations, it so happens that the prime implicants used in the direct approach solution and the prime implicants used in the complementary approach solution together account for *all* of the optional combinations, and there is no optional combination accounted for in both approaches, the minimum sum of products obtained in the direct approach and the minimum product of sums obtained in the complementary approach will be equivalent. If any optional combination is not accounted for in either approach, or is accounted for in both approaches, the two resulting expressions will not be equivalent. The expressions obtained with the two approaches may therefore not be logic equivalents; however, lack of equivalence may occur only for optional combinations.

This discussion on possible lack of equivalence with the two approaches is only of theoretical interest, and is of no concern from a practical standpoint.

The complementary approach will now be applied to the example on page 58. Again, it is suggested that, for practice, the reader solve this problem on his own first.

A B C D	A B C D	A B C D	A B C D
0 0 1 0	0 0 0 0✓	0 0 — 0✓	0 — — 0 *
0 1 1 0		0 — 0 0✓	
0 1 1 1	0 0 1 0✓	— 0 0 0*	
1 0 0 0	0 1 0 0✓		
1 0 0 1	1 0 0 0✓	0 — 1 0✓	
		0 1 — 0✓	
0 0 0 0	0 1 1 0✓	1 0 0 —*	
0 1 0 0	1 0 0 1✓		
		0 1 1 —*	
	0 1 1 1✓		

	0010	0110	0111	1000	1001	
				✓		— 0 0 0
				✓	✓	1 0 0 — *
		✓	✓			0 1 1 — *
	✓	✓				0 — — 0 *

This simple table may be solved intuitively, and it is found that the last three prime implicants are required. The resultant sum of products

$$A\bar{B}\bar{C} + \bar{A}BC + \bar{A}\bar{D}$$

complemented using DeMorgan's theorem, gives the minimum product of sums solution

$$(\bar{A} + B + C)(A + \bar{B} + \bar{C})(A + D)$$

In this example it is seen that the complementary approach involved less work because of the fewer combinations involved.

The two expressions in this example are not logic equivalents since the optional combination 0100 was accounted for in both approaches—by the prime implicant —10— in the direct approach, and by 0— —0 in the complementary approach. When this optional combination $\bar{A}B\bar{C}\bar{D}$ equals 1, the minimum sum of products equals 1, and the minimum product of sums equals 0. For all other possible combinations, the two solutions are equivalent.

$\sum$ and $\prod$ Nomenclature

A compact nomenclature for expressing a Boolean function will be described. In the original table representing the function, consider each row as the represented binary number.[2] The set of decimal equivalents of these numbers, following the symbol "$\sum$," is used to describe the function as a sum of minterms.

EXAMPLE

A	B	C	
0	0	0	0
0	0	1	1
0	1	1	3

$$f(A, B, C) = \sum (0, 1, 3)$$
$$= \bar{A}\bar{B}\bar{C} + \bar{A}\bar{B}C + \bar{A}BC$$

Noting that with n variables, the possible set of numbers ranges from 0 through $2^n - 1$, the complementary sum of minterms is described using the set of numbers missing from the original function. This corresponds

---

[2]The binary number system is discussed in Chapter 6.

to the set of *no-output* combinations. From the previous example, the complementary sum of minterms is represented by

A	B	C	
0	1	0	2
1	0	0	4
1	0	1	5
1	1	0	6
1	1	1	7

$$\bar{f}(A, B, C) = \sum(2, 4, 5, 6, 7)$$
$$= \bar{A}B\bar{C} + A\bar{B}\bar{C} + A\bar{B}C + AB\bar{C} + ABC$$

Remember that complementing a sum of products using DeMorgan's theorem results in a product of sums, and vice versa. Whereas the decimal numbers following "$\sum$" denote the sum of minterms, the decimal numbers following the symbol "$\prod$" denote the complementary product of maxterms.

The original table in the example thus relates to the complementary product of maxterms:

A	B	C	
0	0	0	0
0	0	1	1
0	1	1	3

$$\bar{f}(A, B, C) = \prod(0, 1, 3)$$
$$= (A + B + C)(A + B + \bar{C})(A + \bar{B} + \bar{C})$$

The complementary table relates to the direct product of maxterms:

A	B	C	
0	1	0	2
1	0	0	4
1	0	1	5
1	1	0	6
1	1	1	7

$$f(A, B, C)$$
$$= \prod(2, 4, 5, 6, 7)$$
$$= (A + \bar{B} + C)(\bar{A} + B + C)(\bar{A} + B + \bar{C})(\bar{A} + \bar{B} + C)(\bar{A} + \bar{B} + \bar{C})$$

Thus, given the $\sum$ representation of a sum of minterms, the equivalent $\prod$ representation of the product of maxterms uses the set of numbers missing from the original function, and vice versa.

Optional combinations, with suitable identification, can be incorporated in this nomenclature. The symbols "ϕ" and "d" are commonly used identifiers.

EXAMPLE

A	B	C	
0	0	0	0
0	0	1	1
0	1	1	3
1	0	1	5
1	1	0	6
1	1	1	7

$$f(A, B, C) = \sum (0, 1, 3) + \sum_\phi (5, 6, 7)$$
$$\bar{f}(A, B, C) = \prod (0, 1, 3) + \prod_\phi (5, 6, 7)$$

A	B	C		
0	1	0	2	
1	0	0	4	Complementary table
1	0	1	5	
1	1	0	6	
1	1	1	7	

$$\bar{f}(A, B, C) = \sum (2, 4) + \sum_\phi (5, 6, 7)$$
$$f(A, B, C) = \prod (2, 4) + \prod_\phi (5, 6, 7)$$

The tabular method of minimization can also be used directly when the original expression is in a product of sums form. For example, the product of maxterms

$$(A + \bar{B} + C)(\bar{A} + B + C)(\bar{A} + B + \bar{C})(\bar{A} + \bar{B} + C)(\bar{A} + \bar{B} + \bar{C})$$

can be expressed in the table as

A	B	C
1	0	1
0	1	1
0	1	0
0	0	1
0	0	0

The procedure is the same throughout, the selected prime implicants representing a minimum product of sums.

The reader is cautioned that in some of the literature, the $\prod$ nomenclature is related to this representation. The preceding product of maxterms, in this context, would be written as

$$f(A, B, C) = \prod(0, 1, 2, 3, 5)$$

Compare this with the representation on page 67.

Tabular Method for Multi-Output Functions

The assembly of minimal single-output networks does not assure a minimal multi-output network. Multiple-output networks in which there is more than one function with the same input variables[3] may sometimes be able to share logic blocks in common. For example, if the expression for output-1 is $A\bar{B} + CD$, and the expression for output-2 is $CD + \bar{C}\bar{D}$, the CD term can be shared by both circuits, as shown in Fig.4-1.

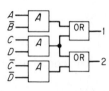

Figure 4-1

The determination of terms that can be shared is not always so obvious. For example, the expressions 1: $\bar{A}BD + \bar{A}C\bar{D}$ and 2: $\bar{A}BC + ACD + \bar{B}C\bar{D}$ (a total of 5 terms and 15 literals) have no term in common; however, an equivalent expression for output-2 is $\bar{A}C\bar{D} + BCD + A\bar{B}C$, which has the term $\bar{A}C\bar{D}$ in common with the output-1 expression (a total now of 4 terms and 12 literals).

Furthermore, a possible common term may not be a prime implicant! Thus, even an examination of all possible equivalent minimum, or, for that matter, irredundant sums of products may not show up possible terms that can be shared to give an optimum multi-output network. As an example, the expressions 1: $\bar{A}\bar{B} + \bar{B}C + \bar{A}C$ and 2: $AC + \bar{B}\bar{D} + BC$ (6 terms, 12 literals) have no terms in common, but the equivalent expressions 1: $\bar{A}\bar{B} + \bar{B}C + \bar{A}BC$ and 2: $AC + \bar{B}\bar{D} + \bar{A}BC$ have the non-prime implicant term $\bar{A}BC$ in common (5 terms, 11 literals).

In this section, the tabular method will be extended to multi-output functions. *Multiple-output prime implicants* are obtained from which is selected a set of sums of products (or products of sums) that is minimum in an overall sense. The method will be illustrated by an example.

The multi-output tables have both input and output columns. All input combinations for which there is at least one output *on* are listed; the outputs that are *on* for each of these input combinations are recorded by a check mark in the appropriate output column. The rows are ordered in the usual manner.

---

[3]At least some of the variables, but not necessarily all, must be the same.

A	B	C	D	1	2	3
0	0	0	1	✓	✓	
0	1	0	1	✓		
0	1	1	0			✓
0	1	1	1	✓	✓	✓
1	0	0	0	✓		✓
1	0	0	1	✓		✓
1	0	1	0	✓		✓
1	0	1	1	✓	✓	✓
1	1	0	1	✓		
1	1	1	1		✓	

A	B	C	D	1	2	3
0	0	0	1	✓	✓	
1	0	0	0	✓		✓
0	1	0	1	✓		
0	1	1	0			✓
1	0	0	1	✓		✓
1	0	1	0	✓		✓
0	1	1	1	✓	✓	✓
1	0	1	1	✓	✓	✓
1	1	0	1	✓		
1	1	1	1		✓	

The combining of the rows is necessarily modified as follows:

1. Only rows with at least one *on* output in common can be combined.
2. In the resulting row, only the *on* outputs that were common are checked.
3. A combining row is "checked off" (signifying that it is not a prime implicant) only if the resulting row accounts for *all* of its *on* outputs.

For some specific examples, note in the table that rows 0001 and 0101

A	B	C	D	1	2	3
0	0	0	1	✓	✓	
0	1	0	1	✓		

cannot be combined because there are no *on* outputs in common.

Rows 0001 and 1001, with *on* output-3 in common, can combine to give

A	B	C	D	1	2	3
0	0	0	1	✓	✓	
1	0	0	1	✓		✓

A	B	C	D	1	2	3
—	0	0	1			✓

the row —001, in which only output-3 is checked. Neither of the combining rows can be checked off, since the resulting row does not account for all of the *on* outputs in either case.

Rows 1000 and 1001, with *on* output-1 and *on* output-3 in common,

A	B	C	D	1	2	3
1	0	0	0	✓		✓
1	0	0	1	✓		✓

A	B	C	D	1	2	3
1	0	0	—	✓		✓

combine to give the row 100—, in which both output-1 and output-3 are checked. Both combining rows are checked off, since the resulting row accounts for all of the *on* outputs in both cases.

Rows 0101 and 0111, with *on* output-1 in common, combine to give

A	B	C	D	1	2	3		
0	1	0	1	✓			✓	
0	1	1	1	✓	✓	✓		

A	B	C	D	1	2	3	
0	1	—	1	✓			

the row 01—1, in which output-1 is checked. The resulting row accounts for all of the *on* outputs in row 0101, but not for all of those in row 0111; therefore, only row 0101 can be checked off. The complete tabulation follows:

A	B	C	D	1	2	3	
0	0	0	1	✓	✓		*
1	0	0	0	✓		✓	✓
0	1	0	1	✓			✓
0	1	1	0		✓		✓
1	0	0	1	✓	✓		✓
1	0	1	0	✓	✓		✓
0	1	1	1	✓	✓	✓	*
1	0	1	1	✓	✓	✓	*
1	1	0	1	✓			✓
1	1	1	1		✓		✓

A	B	C	D	1	2	3	
—	0	0	1			✓	*
1	0	0	—	✓	✓	✓	
1	0	—	0	✓	✓	✓	
0	1	—	1	✓			*
—	1	0	1	✓			*
0	1	1	—		✓		*
1	0	—	1	✓	✓		✓
1	—	0	1	✓			*
1	0	1	—	✓	✓		✓
—	1	1	1		✓		*
1	—	1	1		✓		*

A	B	C	D	1	2	3	
1	0	—	—	✓	✓		*

The multi-output prime implicants having been obtained, we now construct the prime implicant table (see page 72). A column is required for each input-output combination.

For each prime implicant, check marks are placed only in columns corresponding to that prime implicant's *on* outputs. Thus, for example, for prime implicant —001, only output-3 is pertinent; therefore, check marks are placed only in the output-3 0001 and 1001 columns. Check marks are *not* placed in the output-1 1001 or output-2 0001 columns.

The completed table, treated as a whole (i.e., not broken down by output) is solved in the normal manner. The required prime implicants are marked with an asterisk.

One last step must now be made. A selected prime implicant may not be required by *all* of its *on* outputs. Therefore, a check must be made for each

	0001	0111	1011	—001	01—1	—101	011—	1—01	1111	1—11	10——
	*	*				*	*			*	*
3	✓	✓	✓	✓				✓			✓
2	✓	✓	✓						✓	✓	✓
1					✓	✓		✓			✓
3 1011			✓								✓
1010											✓
1001				✓							✓
1000											✓
0111		✓						✓			
0110								✓			
0001	✓			✓							
2 1111									✓	✓	
1011			✓							✓	
0111		✓							✓		
0001	✓										
1 1101						✓		✓			
1011			✓								✓
1010											✓
1001								✓			✓
1000											✓
0111		✓			✓						
0101					✓	✓					

output, to determine if any of its corresponding prime implicants is redundant as far as that particular output is concerned. One case of such redundancy exists in the present example, and relates to output-3. The relevant portion of the table is extracted for instructional purposes. Note that only the selected prime implicants pertinent to output-3 are considered.

			3						1 2 3	
0001	0110	0111	1000	1001	1010	1011				
✓							0 0 0 1	✓ ✓	*
	✓						0 1 1 1	✓ ✓ ✓	*
	✓	✓					0 1 1 —	✓	*
			✓	✓	✓	✓	1 0 — —	✓ ✓	*

Examination of the table shows that, with regard to output-3, prime implicant 0111 is redundant.

Note that this last step does not affect the total number of terms or literals involved; it may, however, reduce the number of OR logic block inputs or, in the event that an expression is reduced to a single term, eliminate an OR logic block.

The optimum set of expressions for the multi-output function is

$$\bar{A}BCD + B\bar{C}D + A\bar{B} \qquad (1)$$

$$\bar{A}\bar{B}\bar{C}D + \bar{A}BCD + ACD \qquad (2)$$

$$\bar{A}\bar{B}\bar{C}D + \bar{A}BC + A\bar{B} \qquad (3)$$

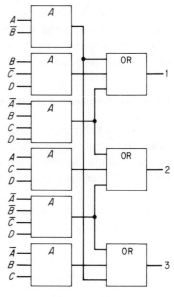

Figure 4-2

The resulting network is shown in Fig 4-2.

If there are any optional input combinations, they are added to the original table in the usual manner; for these combinations, it should be assumed that all outputs are *on*. Also, for a valid input combination, *some* outputs may be optional; it should be assumed that these outputs are *on* also. These optional input-output combinations are used only in obtaining the prime implicants.

In the prime implicant table, there are columns for only the valid input-output combinations.

EXAMPLE

The input combination 0010 can never occur; for the combination 0100, we don't care what any of the outputs are; for the combination 1100, output-1 must be on, output-2 must be off, and output-3 can never occur; for the combination 1110, output-1 and output-3 must be off, and we don't care what output-2 is.

These combinations are added to the original table, as follows:

A	B	C	D	1	2	3
0	0	1	0	✓	✓	✓
0	1	0	0	✓	✓	✓
1	1	0	0	✓		✓
1	1	1	0		✓	

Note that the input combination 1100 with output-1 and output-3 is used for obtaining the prime implicants, whereas the input combination 1100 with output-1 only (which would appear in the original table) is used in the prime implicant table.

Iterative Consensus Method for Obtaining Prime Implicants

In the tabular method described in this chapter, in order to obtain the prime implicants the expanded form had to be obtained, and a large number of terms could be generated and have to be handled in the process. The iterative consensus method for obtaining the prime implicants overcomes these disadvantages.

The method makes use of Theorem 13 in reverse, and Theorems 3 and 11:

$$13. \quad XY + \bar{X}Z = XY + \bar{X}Z + YZ$$
$$3. \quad X + X = X$$
$$11. \quad X + XY = X$$

The method can be stated very simply. Theorem 13 in reverse is applied systematically to all pairs of terms to obtain all possible included terms, which are added to the expression. The pairing continues as the included terms are added. At the same time, terms are eliminated as Theorems 3 and 11 are applied whenever possible. An included term that can be immediately eliminated by the use of Theorems 3 or 11 is not added. The process is continued until no more included terms can be formed, or until the only included terms that can be formed would be immediately eliminated by the use of Theorems 3 or 11. The existing terms at this point comprise all of the prime implicants. (The selection of an optimum set of these prime implicants is then accomplished as described earlier in this chapter.)

EXAMPLE

Find all of the prime implicants from the expression

$$\bar{A}\bar{C}D + A\bar{B}D + AB\bar{C}D + \bar{A}BCD + AB\bar{C}\bar{D}$$

The first term, $\bar{A}\bar{C}D$, paired with the other terms, adds the included terms $\bar{B}\bar{C}D$, $B\bar{C}D$, and $\bar{A}\bar{B}D$; $B\bar{C}D$ eliminates $AB\bar{C}D$, and $\bar{A}\bar{B}D$ eliminates $\bar{A}BCD$. We now have

$$\bar{A}\bar{C}D + A\bar{B}D + AB\bar{C}\bar{D} + \bar{B}\bar{C}D + B\bar{C}D + \bar{A}\bar{B}D$$

The second term, $A\bar{B}D$, with the other terms, adds $A\bar{C}D$ and $\bar{B}D$; $\bar{B}D$ eliminates $A\bar{B}D$, $\bar{B}\bar{C}D$, and $\bar{A}\bar{B}D$. The expression at this point is

$$\bar{A}\bar{C}D + AB\bar{C}\bar{D} + B\bar{C}D + A\bar{C}D + \bar{B}D$$

The next term, $AB\bar{C}\bar{D}$, with the other terms, adds $AB\bar{C}$, which eliminates $AB\bar{C}\bar{D}$. Next, $B\bar{C}D$, with the other terms, adds $\bar{C}D$, which eliminates $\bar{A}\bar{C}D$, $B\bar{C}D$, and $A\bar{C}D$. The expression is now

$$\bar{B}D + AB\bar{C} + \bar{C}D$$

which comprises all of the prime implicants, since there are no more included terms that cannot be immediately eliminated.

This process is more commonly carried out in tabular form. The rules will be restated in this context.

All pairs of rows—original rows, and rows that may be added to the table—are systematically compared for subsumption (Theorems 3 and 11) and consensus (Theorem 13).

A row subsumes another row if it has a 1 in every column in which the other row has a 1, and if it has a 0 in every column in which the other row has a 0. Any row that subsumes another is eliminated. In the following table, the first row subsumes the second and is eliminated:

A	B	C	D
1	0	1	—
—	0	1	—

In the next table, either row is eliminated:

A	B	C	D
1	0	1	—
1	0	1	—

Two rows generate a consensus row if, in one column only, one row has a 1 and the other a 0. The consensus row has a dash in that column. In the remaining columns, the consensus row has a 1 in any column in which either of the two original rows has a 1, it has a 0 in any column in which either of the two original rows has a 0, and it has a dash in any column in which both of the two original rows have dashes. If a consensus row does not subsume any row in the table, it is added to the table. In the following table, the first two rows generate the third (consensus) row. The consensus row is added to the table only if it does not subsume a row already in the table.

A	B	C	D	E	F
0	0	1	0	—	—
1	0	1	—	1	—
—	0	1	0	1	—

The function in the example on page 75 appears in tabular form as

A	B	C	D
0	—	0	1
1	0	—	1
1	1	0	1
0	0	1	1
1	1	0	0

It is suggested that the reader follow through the example in this form.

An orderly implementation of the method is to pair the second row with the first, the third with the first and second, and so forth, pairing each succeeding row with all rows above it. New rows are added to the bottom of the table. The process terminates when the final row has been paired with all rows above it.

As in the tabular method, any optional combinations are included in the table, for prime implicant generation. (Only the valid combinations are accounted for in the selection of an optimum set of prime implicants.)

Iterative Consensus Method for Multi-Output Functions

Application of the iterative consensus method to multiple-output functions requires a modification of the table and an additional rule.

In the consensus method, like the tabular method, the multi-output tables have both input and output columns. The input portion of a row is called the *identifier*; the output portion of a row is called the *tag*. The tag, which can be composed only of dashes and 0's (no 1's), specifies the output functions with which each input term, or identifier, is associated. A dash in an output column indicates that the designated output is associated with the input term in the corresponding row; a 0 indicates that the output is not associated with the term.

Note that the dash/0 nomenclature in the consensus method corresponds, respectively, to the check mark/no check mark nomenclature in the tabular method. The reason for the dash/0 symbology is that the rules for subsumption and consensus remain consistent for the tag. For example, in the following table, the first row subsumes the second and is eliminated:

Identifier				*Tag*		
A	B	C	D	1	2	3
1	0	1	—	0	0	—
—	0	1	—	0	—	—

In the next table, the first two rows generate the third (consensus) row:

A	B	C	D	E	F	1	2	3
0	0	1	0	—	—	0	0	—
1	0	1	—	1	—	0	—	—
—	0	1	0	1	—	0	0	—

One additional rule is necessary for the extension of the consensus method to multi-output functions: Two rows generate an *intersection* (or *product*) row if there is no column in which one row has a 1 and the other a 0. The identifier of the intersection row, like that of a consensus row, has a 1 in any column in which either of the two original rows has a 1, it has a 0 in any column in which either of the two original rows has a 0, and it has a dash in any column in which both of the original rows have dashes. The tag of the intersection row, however, has a dash in any column in which either of the two original rows has a dash, and it has a 0 otherwise. Like a consensus row, if an intersection row does not subsume any row in the table, it is added to the table.

In the following table, the first two rows generate the intersection row 001— —, and the third and fourth rows generate the intersection row 010— —. These two intersection rows are added to the table. (All other generated intersection rows subsume existing rows.)

A	B	C	1	2
0	0	—	0	—
0	—	1	—	0
0	—	0	0	—
0	1	—	—	0
—	0	0	0	—
—	1	1	—	0

Note that a necessary condition for a consensus row to be of value is that the two original rows must have at least one tag column with dashes in both rows; otherwise the consensus row tag will consist of all 0's, implying that the identifier applies to no output. A necessary condition for an intersection row to be of value is that the two original rows must have at least one tag column with a dash in one row and a 0 in the other; otherwise the intersection row will subsume the original rows.

Any optional combinations are added to the table; for these combinations, dashes are entered in all tag columns. Also, for a valid input combination, *some* outputs may be optional; dashes are entered in these tag columns also. These optional input-output combinations are used only in obtaining the prime implicants; only the valid input-output combinations are accounted for in the selection of an optimum set of the prime implicants.

The iterative consensus method for multi-output functions is summarized as follows. All pairs of rows are systematically compared for subsumption, intersection, and consensus. Any row that subsumes another is eliminated. Two rows generate an intersection row if there is no column in which one row has a 1 and the other a 0. Two rows generate a consensus row if, in one column only, one row has a 1 and the other a 0. Each column of a generated row is defined from the two original rows as follows, with one exception:

Original Rows			*Generated Row*
0	1	→	—
—	—	→	—
1	1	→	1
—	1	→	1
0	0	→	0
—	0	→	0

The one exception is the generation of the tag of an intersection row:

$$- \quad 0 \qquad \rightarrow \qquad -$$

Other rules for the iterative consensus method for multi-output functions appear in the literature. The reader is cautioned that these other rules do not assure all of the prime implicants.

PROBLEMS

1. Using the tabular method, obtain the minimum sum of products for $f(A, B, C, D) = \sum(1, 2, 3, 9, 12, 13, 14) + \sum_\phi(0, 7, 10)$.

2. Using the $\prod$ nomenclature, express the function in Problem 1 as a product of maxterms.

3. Using the tabular method, obtain the minimum product of sums for the preceding function.

4. The following table represents the expanded sum of products. The combination $A\bar{B}C\bar{D}$ is optional. Using the tabular method, find all prime implicants, and express algebraically.

A	B	C	D
0	0	0	0
0	0	1	0
0	0	1	1
0	1	0	0
0	1	0	1
0	1	1	1
1	0	0	0
1	0	1	1
1	1	0	1
1	1	1	0
1	0	1	0

5. Given the output combinations and prime implicants below, and using the algebraic method of solution, determine:
   (a) the number of irredundant solutions.
   (b) the minimum-term sum of products.
   (c) the minimum-literal sum of products.

A B C D E				
0 0 1 0 0	0 0 0 1 0	0 1 0 0 1	0 0 0 1 1	
				0 0 — — 0 U
				0 — 0 — 1 V
				0 0 0 — — W
				— 1 — — 1 X
				— — — 1 — Y
				— — 1 — 0 Z

*6. The circuit shown in Fig. 4-3 was designed without the knowledge that the three input combinations $\bar{A}\bar{B}\bar{C}\bar{D}$, $\bar{A}B\bar{C}D$, and $AB\bar{C}\bar{D}$ were invalid. Using the tabular method, redesign the circuit, taking advantage of these optional combinations.

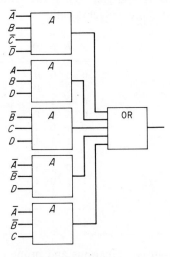

Figure 4-3

7. Using the tabular method, design an optimum multi-output network for the requirements in the following table:

A	B	C	D	1	2	3
0	0	0	0	✓	✓	
0	0	0	1		✓	
0	0	1	0	✓	✓	✓
0	0	1	1			✓
0	1	0	0	✓		✓
0	1	0	1		✓	
0	1	1	1	✓	✓	✓
1	0	0	0	✓	✓	
1	0	1	0	✓	✓	
1	1	1	1	✓		

8. Using the iterative consensus method, find the minimal multi-output network combining the two functions implemented in Fig. 4-4.

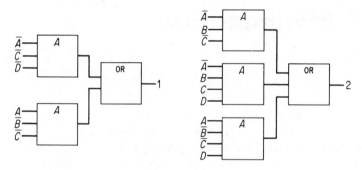

Figure 4-4

5

Map Method of Minimization

The underlying principles of the map method of minimization are basically the same as those for the tabular method. Maps are easy to use because the expression to be simplified is automatically expanded as it is entered on the map, and the prime implicants can be identified by the visual recognition of certain basic patterns.

A map for n variables contains 2^n squares, there being a square on the map for every possible input combination. A 1 is placed in each square representing a combination for which an output is desired; a 0 is placed in each square representing a combination for which no output is desired; and a—is placed in each square representing an optional combination. Often, to reduce the writing, the 0's are omitted, and a blank square is understood to represent a no-output combination.

Figure 5-1 shows two forms of a two-variable map. An analysis of some examples using the two-variable map in both forms will aid in an understanding of the fundamental principles involved.

In entering a map with an expanded term, a 1 is placed in the square of the map corresponding to that expanded term. The entry for $\bar{A}B$ is shown in Fig. 5-2.

In entering the map with a term that is not expanded, a 1 is placed in all squares defined by that term. The entries for the term B are shown in Fig. 5-3.

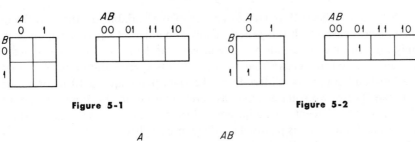

Figure 5-1   Figure 5-2

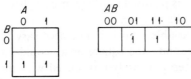

Figure 5-3

Note that if B had first been expanded into $\bar{A}B + AB$, the same two entries would have been made. Thus, B was automatically expanded as it was entered on the map.

In "reading" a map, two 1-squares that are adjacent either horizontally or vertically can be grouped. Larger numbers of 1-squares can also be grouped, the number of squares in a group always being some power of 2; however, for the time being, only groups of two will be considered. The variables that are constant for the group of 1-squares define the group. Thus, the map in Fig.5-3 is read as B.

The map in Fig. 5-3 might have been entered with $\bar{A}B + AB$. With the map entered, it is observed that two 1-squares are adjacent. This group of two 1-squares is defined by B. Therefore, the term B is read from the map, accomplishing the simplification.

The four possible groups of two 1-squares in a two-variable map are shown in Fig. 5-4.

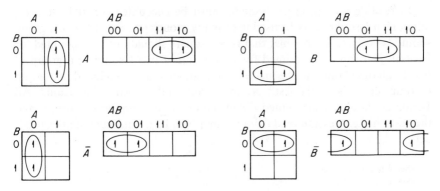

Figure 5-4

Note the "reflected" binary ordering[1] of the variables in the right-hand map in each case. With this ordering, any two adjacent squares will differ in only one variable. Thus, all possible groups can be formed by two adjacent 1-squares. The fourth case, that for $\bar{B}$, warrants special attention. Note that the left-hand square 00 differs from the right-hand square 10 in only one variable. These two squares are considered adjacent in the same sense that the others are adjacent. The adjacency of the two end squares may be better appreciated if the map is pictured as rolled into a cylinder, with the right-hand edge touching the left-hand edge.

While in a two-variable map it is not necessary to get involved in this edge-to-edge wrap-around—the square array could have been used instead —this concept has been purposely introduced with the simple two-variable map because it is used in maps of more variables.

If, instead of the reflected binary ordering, a straight binary ordering[2] had been used, the four previous maps would have looked like Fig. 5-5. Note that the nice relationship of groups always occupying adjacent squares no longer holds. For this reason, the reflected binary ordering is usually used.

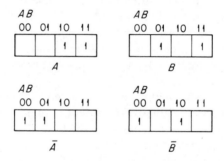

Figure 5-5

In reading a map, every 1-square must be accounted for at least once, although a 1-square may be used in as many groups as desired. Also, a group should be as large as possible; that is, a 1-square should not be accounted for by itself if it can be accounted for in a group of two 1-squares; a group of two 1-squares should not be made if the 1-squares can be included in a group of four; etc. These "largest" groups correspond to prime implicants. All 1-squares should be accounted for in the minimum number of groups, and the resulting expression read from the map will be a minimum sum of products.

[1]See Chapters 6 and 7.
[2]See Chapters 6 and 7.

EXAMPLE

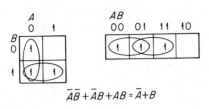

$$\overline{A}\overline{B} + \overline{A}B + AB = \overline{A} + B$$

Figure 5-6

In this example, there are two groups of two 1-squares each, one group defined by $\overline{A}$ and the other group by B (Fig. 5-6). Note that the combination $\overline{A}B$ was used in both groups.

EXAMPLE

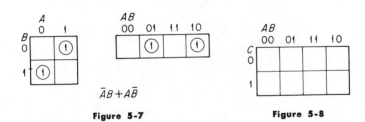

$$\overline{A}B + A\overline{B}$$

Figure 5-7   **Figure 5-8**

In the example of Fig. 5-7 there is no simplification possible; the two 1-squares are not adjacent horizontally or vertically, and $\overline{A}B + A\overline{B}$ is a minimum expression.

The construction of a three-variable map is shown in Fig. 5-8. Figure 5-9 shows, for study, some familiar examples of groups of two 1-squares on three-variable maps.

Groups of four 1-squares may occur either in a straight line array or in a square array, as shown in Fig. 5-10. Note again the concept of the edge-to-edge wrap-around in the $\overline{B}$ example of Fig. 5-10.

Some additional examples are given in Figs. 5-11 to 5-14.

In Fig. 5-12, there are two equally good solutions; the $\overline{A}BC$ square can be accounted for by either $\overline{A}C$ or BC.

Note, in all of these examples, that all 1-squares have been accounted for at least once and that all groups are as large as possible.

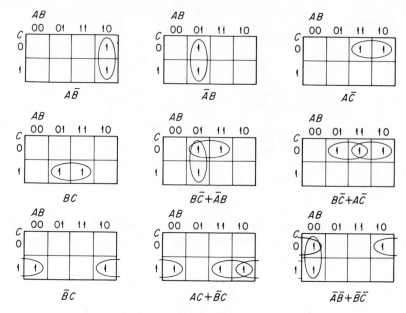

Figure 5-9

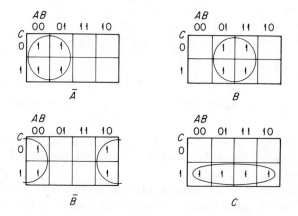

Figure 5-10

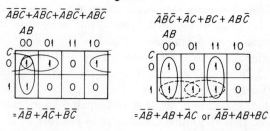

Figure 5-11   Figure 5-12

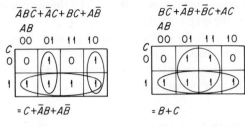

$\overline{A}B\overline{C}+\overline{A}C+BC+A\overline{B}$

$B\overline{C}+\overline{A}B+\overline{B}C+AC$

= $C+\overline{A}B+A\overline{B}$

= $B+C$

Figure 5-13   **Figure 5-14**

Complementary Approach with Map Method

As in the tabular method, it is possible with maps to use the complementary approach. In the preceding four examples, 0's have purposely been entered on the maps in preparation for the discussion of the complementary approach.

In the complementary approach, the 0-squares, rather than the 1-squares, are grouped. Since the resultant sum of products is the complement of the desired expression, this sum of products is complemented using DeMorgan's theorem. The final expression is thus in a minimum product of sums form.

The complementary approach to the preceding four examples is shown in Fig. 5-15.

In the complementary approach, the minimum product of sums can be read directly from the map by the mental application of DeMorgan's theorem during the process of reading. For instance, in the second example, instead of the 010 entry being read as the product $\overline{A}B\overline{C}$, and later complemented, it

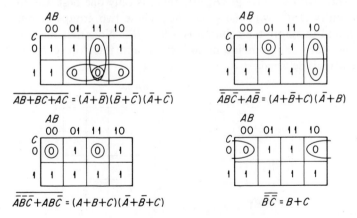

$\overline{AB+BC+AC} = (\overline{A}+\overline{B})(\overline{B}+\overline{C})(\overline{A}+\overline{C})$

$\overline{A}\overline{B}\overline{C}+A\overline{B} = (A+\overline{B}+C)(\overline{A}+B)$

$\overline{A}\overline{B}\overline{C}+AB\overline{C} = (A+B+C)(\overline{A}+\overline{B}+C)$

$\overline{B}\overline{C} = B+C$

Figure 5-15

can be read directly as the sum $(A + \bar{B} + C)$ by the mental complementation of the variables as the map is read.

Maps are convenient for converting an expression from the sum of products form to the product of sums, or vice versa. For example, Fig. 5-15 illustrates sums of products entered on a map, with 1's, and products of sums read from the map by grouping the 0's.

"Method of Attack"

The approach in reading the optimum solution from a map is to account first for all 1-squares that can be grouped in only one best way, that is, 1-squares that are contained in essential prime implicants, leaving until last those in which a choice of prime implicants is involved. Any 1-squares that do not combine with any others are accounted for by themselves. Any 1-squares that combine with only one other 1-square are accounted for in such groups of two. If a 1-square combines with exactly two other squares, look to see if there is a fourth 1-square that completes a group of four. If there is, the four entries are accounted for as a group; if not, then there is a choice involved as to which of the two groups of two to choose, and such decisions should be left until last. And so forth. Following the establishment of all essential prime implicants, any remaining 1-squares should be combined into the fewest possible groups.

The following simple example illustrates the approach suggested. This example is of interest also because it illustrates the included term theorem.

EXAMPLE

In this example (Fig. 5-16), there is only one best way to account for the entry $\bar{A}\bar{B}C$: with the group $\bar{A}\bar{B}$; and there is only one best way to account for the entry $AB\bar{C}$: with the group $B\bar{C}$. These two groups account for all entries, and therefore the solution is $\bar{A}\bar{B} + B\bar{C}$. Note that the entries $\bar{A}\bar{B}\bar{C}$ and $\bar{A}B\bar{C}$ each can combine in two ways, and therefore consideration of these entries is deferred.

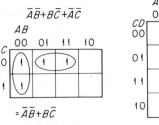

Figure 5-16

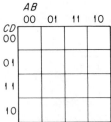

Figure 5-17

A four-variable map is shown in Fig. 5-17. Note the reflected ordering in both the horizontal and vertical directions. In four-variable maps, not only are the left and right edges adjacent, but the top and bottom edges are also adjacent. A mental picture of this left-right top-bottom wrap-around can be formed if one considers the map as rolled into a cylinder with the left and right edges touching, and then the cylinder rolled into a torus with the top and bottom edges touching.

Figure 5-18 shows a few examples of groups involving these wrap-around adjacencies.

Figure 5-19 shows two examples of groups of eight 1-squares.

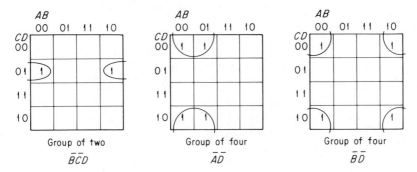

Figure 5-18

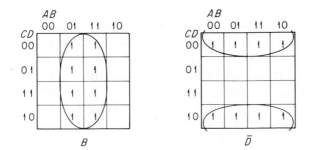

Figure 5-19

Groups should always be as large as possible; that is, every group should correspond to a prime implicant. However, a group should not be made just because it is large. The following example illustrates this important point.

EXAMPLE

In Fig. 5-20, the group $\bar{A}\bar{C}$ looks very attractive because it is the only group of four on the map. However, all of the entries in this group can

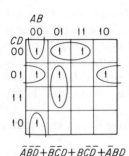

$$\bar{A}\bar{B}\bar{D}+\bar{B}\bar{C}D+\bar{B}\bar{C}\bar{D}+\bar{A}BD$$

Figure 5-20

combine in more than one way and it is best to consider first those entries that cannot combine in more than one way. Study of the map reveals that each of the other four 1-squares combines with only one other 1-square. When these four groups of two have been made, it is found that every 1-square on the map has been accounted for; thus, the term $\bar{A}\bar{C}$ is redundant.

Figures 5-21 and 5-22 are given for study in both entering and reading a map. In Fig. 5-21, the groups are numbered to correspond to the terms from which the map was entered.

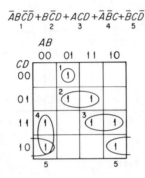

Figure 5-21

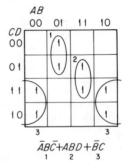

Figure 5-22

A note about entering a map. As each literal in a term is considered, the already defined portion of the map is halved. For example, entering term 3, ACD, in Fig. 5-21, the literal A defines the right half of the map, the literal C defines the lower half of this area, and of this portion of the map, the literal D defines the upper half. The map is repeated in Fig. 5-22, showing the groups that are read. These groups are numbered to correspond to the terms in the final expression.

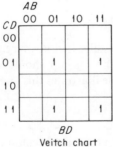

Veitch chart
Straight binary ordering

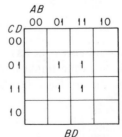

Karnaugh map
Reflected binary ordering

Figure 5-23

Another comparison of the straight binary ordering and the reflected binary ordering, this time in four-variable maps, is shown in Fig. 5-23. The first application of this type of graphical approach to simplification is accredited to E. W. Veitch. The Veitch chart used the straight binary ordering shown on the left. M. Karnaugh modified the Veitch chart, using the reflected binary ordering shown on the right. The resulting improvement is that, in the Karnaugh map, all groups are adjacent rather than some of them being scattered as in the Veitch chart. An alternative method of labeling the Karnaugh map is shown in Fig. 5-24.

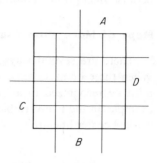

Figure 5-24

Optional Combinations with Map Method

Optional combinations are entered on a map by placing —'s[3] in the corresponding squares. These optional entries may be used, with the 1-squares, in accounting for all of the 1-squares in fewer and/or larger groups. Note that any optional entries *may* be accounted for, but only the 1-squares *must* be accounted for.

In Fig. 5-25, optional combinations are used to advantage. The optional entries $\bar{A}B\bar{C}\bar{D}$ and $\bar{A}B\bar{C}D$ are used with the 1-squares $\bar{A}\bar{B}\bar{C}\bar{D}$ and $\bar{A}\bar{B}\bar{C}D$ to give the group $\bar{A}\bar{C}$; the optional entries $AB\bar{C}D$ and $ABC\bar{D}$ are used with the 1-squares $ABCD$ and $A\bar{B}C\bar{D}$ to give the group AC. The optional entry $AB\bar{C}\bar{D}$ is not used.

Optional combinations can be used in both the direct and complementary approach, as shown in Fig.

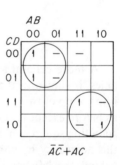

$$\bar{A}\bar{C}+AC$$

Figure 5-25

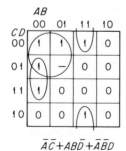

$$\bar{A}\bar{C}+AB\bar{D}+\bar{A}BD$$

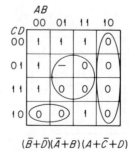

$$(\bar{B}+\bar{D})(\bar{A}+B)(A+\bar{C}+D)$$

Figure 5-26

[3]Other symbols, such as ϕ and d, are also used.

5-26. (Note that the two resultant expressions are not true equivalents because the optional combination was used in a group in both approaches.)

Maps of More than Four Variables

There are several ways of drawing maps of more than four variables. When three or more variables are involved *in one dimension*, new patterns must be recognized in addition to the adjacencies already discussed. For example, in the five-variable map of Fig. 5-27, squares equidistant from the vertical center line are considered adjacent.

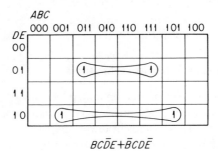

$$BC\bar{D}E + \bar{B}CD\bar{E}$$

Figure 5-27

A more general approach, that can be extended to any number of variables, is shown in the five-variable map of Fig. 5-28. This five-variable map is made up of two four-variable maps drawn side by side. Groups are formed as before except that, in addition to the adjacencies already discussed, corresponding squares on the two maps are considered adjacent. One may picture this adjacency by considering the right-hand map as being situated directly behind the left-hand map, making a three-dimensional map four squares across by four squares down by two squares deep.

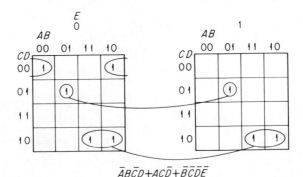

$$\bar{A}B\bar{C}D + A\bar{C}\bar{D} + \bar{B}\bar{C}\bar{D}\bar{E}$$

Figure 5-28

In Fig. 5-28, the $\bar{A}B\bar{C}D\bar{E}$ entry on the left-hand map is adjacent to the $\bar{A}B\bar{C}DE$ entry on the right-hand map, giving the group $\bar{A}B\bar{C}D$. The $AC\bar{D}$ group is also made up of 1-squares from both maps. The $\bar{B}\bar{C}\bar{D}\bar{E}$ group is made up of 1-squares from the $\bar{E}$ map only.

A six-variable map is made up of four four-variable maps drawn in a square array. In addition to the groups that can be formed on any one four-variable map, groups can also be made from corresponding entries on maps horizontally or vertically adjacent.

In the map of Fig. 5-29, the group $\bar{A}B\bar{C}D\bar{E}$ comes from corresponding entries on the two left-hand maps; the group $AB\bar{C}\bar{D}\bar{F}$ comes from corresponding entries on the two upper maps; the term $A\bar{B}CD$ comes from corresponding entries on all four maps.

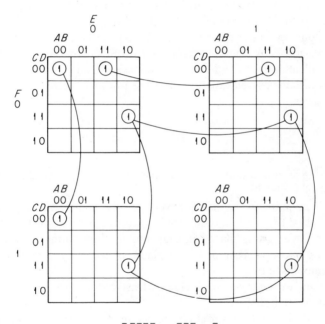

$$\bar{A}B\bar{C}D\bar{E}+AB\bar{C}\bar{D}F+A\bar{B}CD$$

Figure 5-29

A seven-variable map is made by placing two six-variable maps side by side. In addition to the groups that can be made on each six-variable map, groups can also be formed from corresponding entries on the two maps. An eight-variable map is made by placing four six-variable maps in a square array; a nine-variable map is made by placing two eight-variable maps side by side; a ten-variable map is made by placing four eight-variable maps in a square array; and so forth.

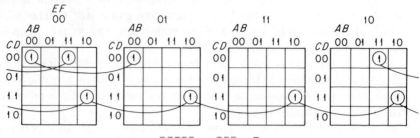

$$\overline{ABC}DE+\overline{ABC}DF+\overline{AB}\overline{C}D$$

Figure 5-30

Sometimes a six-variable map is drawn as in Fig. 5-30. The four four-variable maps can be pictured as being placed one behind the other, the left-hand map on top and the right-hand map on the bottom, forming a cube four squares across by four squares down by four squares deep. In this cube, the left-hand and right-hand faces are considered adjacent, the front and back faces are considered adjacent, and the top and bottom faces are considered adjacent.

Summary

Following is a summary of some pertinent points regarding the map method of minimization.

There is a square on the map for every possible combination of variables.

A 1 is placed in each square representing a combination for which an output is desired; a 0 is placed in each square representing a combination for which no output is desired; a — is placed in each square representing an optional combination.

Each 1-square *must* be considered at least once. Each 1-square *may* be considered as often as desired.

Generally, all 1-squares should be accounted for in the minimum number of groups.

Each group should be as large as possible; that is, each group should correspond to a prime implicant.

The number of squares in a group must always be some power of 2. In a group of 2^m squares, m variables will occur in all possible combinations. If the total number of variables is n, then $(n - m)$ variables will be constant in these 2^m squares, and these $(n - m)$ variables will define the group.

If the groups are made in an optimum manner, the expression read from the map will be a minimum sum of products.

The complementary approach may be used, in which case the 0-squares rather than the 1-squares are grouped. The groups are complemented and a minimum product of sums is obtained.

Optional combinations may be used to obtain fewer and/or larger groups.

The map method, like the tabular method, can also be directly adapted to the minimization of expressions in the product of sums form, each 1-square representing an expanded sum, and the selected groups representing a minimum product of sums.

Broadly speaking, relatively simple functions are minimized by the application of the Boolean theorems. For more complex functions, the map method is used. For functions beyond the capacity of the map method, the tabular method is used.

Whether the map method can be used successfully depends not only on the number of variables involved but also on whether or not enough of the 1-squares can be accounted for by essential prime implicants. A seven-variable function may yield a map with too many 1-squares that are contained in more than one prime implicant, the map being too unwieldy to solve. On the other hand, a ten-variable function may yield a map in which most of the 1-squares are contained in essential prime implicants, and one can be sure that the optimum solution has been obtained.

On larger maps that are not workable because of too many nonessential prime implicants, the map method and tabular method can be combined. After all essential prime implicants have been found with the map, all of the remaining 1-squares and all corresponding nonessential prime implicants are entered on the prime implicant table and the remainder of the problem is solved in this manner.

Map Method for Multi-Output Functions

Maps can be used in the optimization of multi-output functions. The method will be illustrated using the multi-output function on page 70, which was used as an example with the tabular method. The reader can benefit by comparing, step by step, the tabular and map approaches with this example.

Refer to Fig. 5-31. First, a map is drawn for each individual function. Then a map is drawn for each pair of functions. A 1-square appears in a two-function map only if there are corresponding 1-squares in both single-function maps; that is, the two-function map is the intersection, or product, of the two corresponding single-function maps.

In general, a map is drawn for each combination of three functions, four functions, and so forth, a map of all of the functions finally completing the set. Each multifunction map is an intersection of all of the corresponding

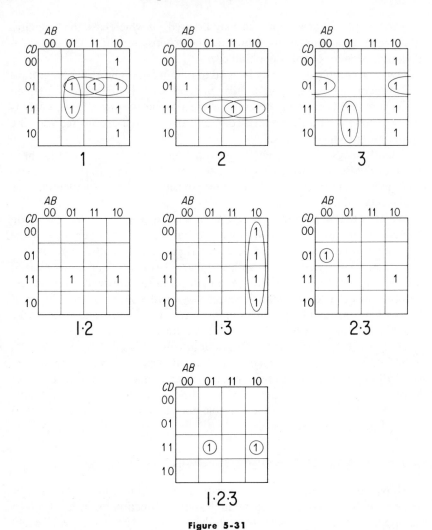

Figure 5-31

single-function maps. In the example, three two-function maps and one three-function map complete the set.

Now, starting with the map of all of the functions and working back to the single-function maps, identify all of the prime implicants in each map, subject to the condition that if there is a prime implicant in a map of certain functions, that same prime implicant will not be identified in any map of a subset of those functions. In the example in Fig. 5-31, all of the prime implicants are shown grouped for identification. Note that the two prime implicants in the 1·2·3 map are not identified in any of the other maps; the prime implicant 10— — in the 1·3 map is not identified in the 1 map or the 3 map; and the prime implicant 0001 in the 2·3 map is not identified in the 2 map. It is important to realize that here the groupings represent all of the prime implicants, not just those that enter into the solution.

Depending on the complexity of the problem, the optimum set of prime implicants can now be selected directly from the map, or the information can be transferred to a prime implicant table and the selection made by that means. The reader may verify that the prime implicant table on page 72 would result from Fig. 5-31.

For any optional input combinations, dashes are entered on all maps. If, for a valid input combination, *some* outputs are optional, dashes are entered in the corresponding squares of the single-function maps for those outputs. For the intersection of single-map entries that are all dashes, a dash entry is made in the multi-function map; for the intersection of single-map entries that consist of 1's and dashes, a 1 entry is made in the multi-function map.

PROBLEMS

Minimize the following expressions using the map method.

1. $\bar{A}\bar{B}C + AD + \bar{D}(B + C) + A\bar{C} + \bar{A}\bar{D}$

***2.** $ACD + B\bar{D} + \bar{B}\bar{C} + \bar{B}D + \bar{C}\bar{D}$
Optional combinations: $\bar{A}\bar{B}C\bar{D}$, $A\bar{B}C\bar{D}$, $\bar{A}B\bar{C}D$

3. $B\bar{C} + \bar{A}B + BC\bar{D} + \bar{A}\bar{B}D + A\bar{B}\bar{C}D$

4. $B\bar{C}\bar{D} + \bar{A}\bar{B}D + A\bar{C}D + \bar{A}BC + \bar{A}B\bar{C}D$
Optional combinations: $\bar{A}\bar{B}\bar{C}\bar{D}$, $ABCD$, $A\bar{B}C\bar{D}$

***5.** $C(B\bar{D} + \bar{B}D) + \bar{A}C(B + D) + CD(A + B) + AD(\bar{B} + \bar{C}) + AB$

6. Read the map in Fig.5-32.

E = 0

CD \ AB	00	01	11	10
00	1	—		1
01	—	1	1	—
11			1	
10	1			1

E = 1

CD \ AB	00	01	11	10
00		—		1
01		1		
11				
10	1			1

Figure 5-32

7. Read the map in Fig. 5-32 using the complementary approach.

8. Using the map method, obtain the minimum sum of products:

$$\bar{A}\bar{B}\bar{C}D\bar{E}\bar{F} + \bar{A}BCD\bar{E}\bar{F} + A\bar{B}\bar{D} + BC\bar{D}\bar{E}F + B\bar{C}\bar{D}E\bar{F}$$
$$+ \bar{A}B\bar{C}\bar{D}E + \bar{A}\bar{B}C\bar{D}F + \bar{A}CDE\bar{F} + \bar{A}\bar{C}D\bar{E}F$$

Optional combinations: $\bar{A}B\bar{C}D\bar{E}\bar{F}, \bar{A}BCD\bar{E}\bar{F}$

*9. Using the map method, obtain the minimum sum of products:

$$AB\bar{C}D\bar{E}\bar{F} + AB\bar{D}\bar{E}F + AB\bar{C}\bar{D}F + A\bar{C}DE\bar{F} + \bar{A}B\bar{D}E F$$
$$+ \bar{A}\bar{B}\bar{C}D\bar{E}\bar{F} + \bar{A}\bar{B}\bar{C}\bar{D}E + \bar{A}\bar{C}D\bar{E}\bar{F} + \bar{B}C\bar{D}$$

Optional combinations: $A\bar{B}\bar{C}D\bar{E}\bar{F}, \bar{A}B\bar{C}D\bar{E}\bar{F}$

6

Number Systems

Number Systems

A number such as

$$2{,}547.16$$

is not normally thought of as being composed of two 1,000's, five 100's, four 10's, seven 1's, one $\frac{1}{10}$, and six $\frac{1}{100}$'s. However, in a discussion of number systems in general, it will be helpful to think of numbers broken down in this way.

In general, the right-most digit to the left of the radical point represents the number of 1's or B^0s, where B is the *base* or *radix* of the number system. The next digit to the left represents the number of B^1s; the next digit to the left represents the number of B^2s; the next digit to the left represents the number of B^3s; etc.

The left-most digit to the right of the radical point represents the number of B^{-1}s; the next digit to the right represents the number of B^{-2}s; the next digit to the right represents the number of B^{-3}s; etc.

In the decimal number system, the base is 10. The analysis for the decimal number 2,547.16 is shown below:

2	5	4	7	1	6
$= 2 \times 10^3$	5×10^2	4×10^1	7×10^0	1×10^{-1}	6×10^{-2}
$= 2 \times 1000$	5×100	4×10	7×1	$1 \times \frac{1}{10}$	$6 \times \frac{1}{100}$

$$
\begin{aligned}
2 \times 1000 &= 2000 \\
5 \times 100 &= 500 \\
4 \times 10 &= 40 \\
7 \times 1 &= 7 \\
1 \times \tfrac{1}{10} &= \tfrac{1}{10} \\
6 \times \tfrac{1}{100} &= \tfrac{6}{100} \\
\hline
2547\tfrac{16}{100} &= 2547.16
\end{aligned}
$$

In a number system to the base B, there are B different symbols, ranging from 0 to $B - 1$. Thus, in the decimal system, base 10, there are ten different symbols, 0 through 9.

The preceding general concepts will now be applied to number systems with other bases. The following number is written in the base 8, or octal, number system:

$$256.71$$

This number is *not* two hundred fifty-six and seventy-one hundredths. Since it is written in the base 8 number system, it is analyzed as follows:

$$
\begin{array}{ccccccc}
2 & 5 & 6 & 7 & 1 & & \text{(base 8)} \\
= 2 \times 8^2 & 5 \times 8^1 & 6 \times 8^0 & 7 \times 8^{-1} & 1 \times 8^{-2} \\
= 2 \times 64 & 5 \times 8 & 6 \times 1 & 7 \times \tfrac{1}{8} & 1 \times \tfrac{1}{64}
\end{array}
$$

$$
\begin{aligned}
2 \times 64 &= 128 \\
5 \times 8 &= 40 \\
6 \times 1 &= 6 \\
7 \times \tfrac{1}{8} &= \tfrac{7}{8} \\
1 \times \tfrac{1}{64} &= \tfrac{1}{64} \\
\hline
174\tfrac{57}{64} & \quad \text{(base 10)}
\end{aligned}
$$

256.71 in the base 8 number system represents two 64's, five 8's, six 1's, seven $\tfrac{1}{8}$'s and one $\tfrac{1}{64}$. The total $174\tfrac{57}{64}$ in the base 10 or decimal system, then, is the equivalent of 256.71 in base 8.

The above type of analysis can be used to convert from a number in any base to its decimal equivalent. A convenient method for converting from a decimal number to a number in some other base follows. The decimal number is separated into two parts: that to the left of the decimal point, and that to the right of the decimal point. Each part is handled in a different way.

The left part of the number is repeatedly divided by the base to which it is to be converted, and the remainders are recorded for each division. This procedure is continued until the quotient 0 is reached. The remainders, reading from the last remainder to the first, represent the left part of the number in the new base.

The right part of the number is repeatedly multiplied by the base to which it is to be converted, and the carries are recorded for each multiplication. This procedure is continued until the product 0 is reached, or until the desired number of places is obtained. The carries, reading from the first carry to the last, represent the right part of the number in the new base.

As an example, the decimal number 174.890625 will be converted to its equivalent in the base 8:

Left part:		Right part:
Remainder		*Carry*

$$
\begin{array}{lll}
8 & |174 & 6 \\
8 & |\,21 & 5 \\
8 & |\;2 & 2 \\
& \;\;0 &
\end{array}
\qquad
\begin{array}{ll}
& .890625 \\
& \times\,8 \\
7 & .125000 \\
& \times\,8 \\
1 & .000000
\end{array}
$$

174 (base 10) = 256 (base 8)   .890625 (base 10) = .71 (base 8)

174.890625 (base 10) = 256.71 (base 8)

For an example of an approximate conversion, the decimal number .14159 will be converted to its "equivalent" in base 3 (ternary), correct to four places:

Carry

$$
\begin{array}{ll}
& .14159 \\
& \times\,3 \\
0 & .42477 \\
& \times\,3 \\
1 & .27431 \\
& \times\,3 \\
0 & .82293 \\
& \times\,3 \\
2 & .46879
\end{array}
$$

.14159 (base 10) $\approx$ .0102 (base 3)

A few more examples of other number systems and their conversion to and from decimal are given below for study.

(a)   142 (base 5) = ? (base 10)

$$
\begin{array}{ccc}
1 & 4 & 2 \\
= 1 \times 5^2 & 4 \times 5^1 & 2 \times 5^0 \\
= 1 \times 25 & 4 \times 5 & 2 \times 1
\end{array}
$$

$$
\begin{aligned}
2 \times 1 &= 2 \\
4 \times 5 &= 20 \\
1 \times 25 &= 25 \\
\hline
&\;\;47
\end{aligned}
$$

142 (base 5) = 47 (base 10)

(b)   47 (base 10) = ? (base 5)

$$
\begin{array}{lll}
5 & |47 & 2 \\
5 & |\;9 & 4 \\
5 & |\;1 & 1 \\
& \;\;0 &
\end{array}
$$

47 (base 10) = 142 (base 5)

(c)   201 (base 3) = ? (base 10)

$$
\begin{array}{ccc}
2 & 0 & 1 \\
= 2 \times 3^2 & 0 \times 3^1 & 1 \times 3^0 \\
= 2 \times 9 & 0 \times 3 & 1 \times 1
\end{array}
$$

$$
\begin{array}{rcl}
1 \times 1 &=& 1 \\
0 \times 3 &=& 0 \\
2 \times 9 &=& \underline{18} \\
& & 19
\end{array}
$$

201 (base 3) = 19 (base 10)

(d)   19 (base 10) = ? (base 3)

$$
\begin{array}{l}
3 \ \underline{|19} \ \ 1 \\
3 \ \ \underline{|6} \ \ \ 0 \\
3 \ \ \underline{|2} \ \ \ 2 \\
\ \ \ \ \ 0
\end{array}
$$

19 (base 10) = 201 (base 3)

Question: What is the decimal equivalent of 182 (base 8)?

Answer: There can be no such number as 182 in the base 8; in this base there are only eight allowable symbols, 0 through 7. There is no symbol for 8 in the base 8 number system, any more than there is a symbol for 10 in the decimal system. In the base 8 number system, a 1 in the 8's position represents the value 8, just as in the decimal system, a 1 in the 10's position represents the value 10.

The binary or base 2 number system is of particular importance in computers. Each position in the binary number system has only two possible symbols, 0 or 1. Therefore, binary arithmetic or the storage of binary numbers is a "natural" for circuits which have only two possible states.

Below is an example of binary-to-decimal and decimal-to-binary conversion:

101011 (base 2) = ? (base 10)

$$
\begin{array}{cccccc}
1 & 0 & 1 & 0 & 1 & 1 \\
= 1 \times 2^5 & 0 \times 2^4 & 1 \times 2^3 & 0 \times 2^2 & 1 \times 2^1 & 1 \times 2^0 \\
= 1 \times 32 & 0 \times 16 & 1 \times 8 & 0 \times 4 & 1 \times 2 & 1 \times 1
\end{array}
$$

$$
\begin{array}{rcl}
1 \times 1 &=& 1 \\
1 \times 2 &=& 2 \\
0 \times 4 &=& 0 \\
1 \times 8 &=& 8 \\
0 \times 16 &=& 0 \\
1 \times 32 &=& \underline{32} \\
& & 43
\end{array}
$$

101011 (base 2) = 43 (base 10)

43 (base 10) = ? (base 2)

```
2 |43  1
2 |21  1
2 |10  0
2 | 5  1
2 | 2  0
2 | 1  1
     0
```

43 (base 10) = 101011 (base 2)

Following is a table of the values 0 to 20 in various number systems:

Number Systems

Decimal, Base 10	Binary, Base 2	Ternary, Base 3	Quinary, Base 5	Octal, Base 8	Hexadecimal, Base 16
0	0	0	0	0	0
1	1	1	1	1	1
2	10	2	2	2	2
3	11	10	3	3	3
4	100	11	4	4	4
5	101	12	10	5	5
6	110	20	11	6	6
7	111	21	12	7	7
8	1000	22	13	10	8
9	1001	100	14	11	9
10	1010	101	20	12	A
11	1011	102	21	13	B
12	1100	110	22	14	C
13	1101	111	23	15	D
14	1110	112	24	16	E
15	1111	120	30	17	F
16	10000	121	31	20	10
17	10001	122	32	21	11
18	10010	200	33	22	12
19	10011	201	34	23	13
20	10100	202	40	24	14

Note that in the base 16 number system, sixteen symbols are needed; in addition to 0 through 9, the symbols A through F are commonly used.

Although there are methods for directly converting from one base to another, neither base being decimal, it is generally convenient to perform the conversion in two steps: from the original base to decimal, and from decimal to the new base.

An exception is when one base is a power of the other, in which case the conversion is simple. Denote the lower-valued base as B_x, the higher-valued base as B_y, and their relationship as $B_x^n = B_y$. To convert from B_x to B_y, start from the radical point and, both to the left and right, divide the digits in the B_x number into groups of n. Write each group as a B_y digit, and the conversion is accomplished.

The representation of a binary number in the more compact octal ($2^3 = 8$; $n = 3$) or hexadecimal ($2^4 = 16$; $n = 4$) is of particular usefulness. As an example, the binary number 11100.01101 converted to octal is

$$011 \mid 100.011 \mid 010$$
$$3 \mid 4 \ . \ 3 \mid 2$$

and converted to hexadecimal is

$$0001 \mid 1100.0110 \mid 1000$$
$$1 \mid C \ . \ 6 \mid 8$$

The method of conversion from B_y to B_x should be obvious.

Binary Adders

The purpose of this section is to illustrate how switching circuits can be used to perform binary addition.

In the arithmetic addition of two binary digits or "bits," there are four

$$
\begin{array}{c}
A \\
+B \\
\hline
\text{Sum}
\end{array}
\quad
\begin{array}{c}
0 \\
0 \\
\hline
0
\end{array}
\quad
\begin{array}{c}
0 \\
1 \\
\hline
1
\end{array}
\quad
\begin{array}{c}
1 \\
0 \\
\hline
1
\end{array}
\quad
\begin{array}{c}
1 \\
1 \\
\hline
0
\end{array}
$$

Carry-out

Carry-out into next higher-order position

Figure 6-1

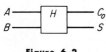

Figure 6-2

possible combinations, as shown in Fig. 6-1. A circuit for adding two bits is called a *half-adder* (Fig. 6-2). A half-adder has two inputs, for the two bits to be added, and two outputs, one for the sum, S, and one for the "carryout," C_0, into the next higher-order position.

Observation of the four possible combinations of two bits shows that the sum equals 1 only when $A = 1$ and $B = 0$, or when $A = 0$ and $B = 1$. Furthermore, there is a carry-out into the next higher-order position only when A and B both equal 1.

Boolean expressions for the sum and carry-out outputs of the half-adder can be written as follows:

$$S = A\bar{B} + \bar{A}B$$
$$C_0 = AB$$

The logic circuit to implement these expressions is shown in Fig. 6-3. This circuit can be simplified by the manipulation of the Boolean expression for the sum:

$$S = A\bar{B} + \bar{A}B$$
$$= (A + B)(\bar{A} + \bar{B})$$
$$= (A + B)(\overline{AB})$$

The simplified half-adder is shown in Fig. 6-4.

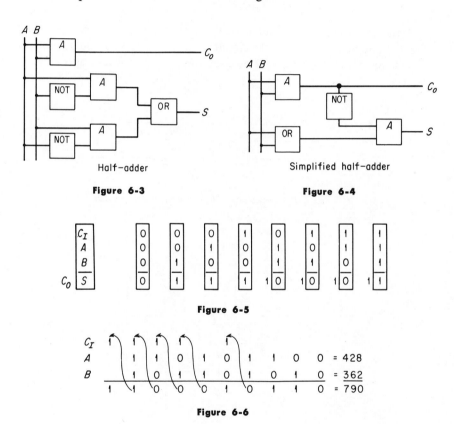

Half-adder

Figure 6-3

Simplified half-adder

Figure 6-4

Figure 6-5

Figure 6-6

When two bits, A and B, are added in a position, and there is a "carry-in," C_I, from the next lower-order position, three bits in all must be added. In the addition of three bits, there are eight possible combinations (Fig. 6-5). Figure 6-6 shows an example of binary addition. A device for adding three bits is called a *full-adder* (Fig. 6-7). A full-adder has three inputs: A, B, and C_I; and two outputs: S and C_0.

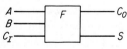

Figure 6-7

In the addition of three bits, the sum equals 1 only when exactly one or three of the bits equal 1. The sum can be expressed as

$$S = A\bar{B}\bar{C}_I + \bar{A}B\bar{C}_I + \bar{A}\bar{B}C_I + ABC_I$$

The carry-out into the next higher-order position equals 1 only when exactly two or three of the bits equal 1. The carry-out can be expressed as

$$C_0 = AB\bar{C}_I + A\bar{B}C_I + \bar{A}BC_I + ABC_I$$
$$= AB + AC_I + BC_I$$

A full-adder can be constructed with two half-adders and an OR circuit (Fig. 6-8). Some key points in the circuit have been defined to aid in analysis.

The general structure of a 4-position binary adder would appear as in Fig. 6-9.

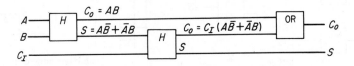

$$S = (A\bar{B} + \bar{A}B)\,\bar{C}_I + (\bar{A}\bar{B} + AB)\,C_I = A\bar{B}\bar{C}_I + \bar{A}B\bar{C}_I + \bar{A}\bar{B}C_I + ABC_I$$
$$C_0 = AB + C_I\,(A\bar{B} + \bar{A}B) = AB + AC_I + BC_I$$

Figure 6-8

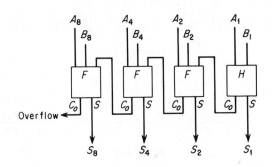

Figure 6-9

Binary-Coded-Decimal Adder

A binary-coded-decimal (BCD) adder will now be discussed. The BCD code differs from the straight binary number representation in that in the BCD code *each decimal digit is binary-coded.* For example, the decimal number 13, in straight binary number representation, is

1101

whereas in the BCD code, 13 is represented by

0001   0011

the decimal digits 1 and 3 each being binary-coded.

In the BCD code, the highest allowable binary representation is 1001 (9). Therefore, the highest two numbers that may be added are 1001 + 1001 (9 + 9). Also, there may be a carry-in from the next lower-order position. Thus, the maximum sum that can occur is 10011 (19). However, when the sum exceeds 1001 (9), a correction must be made, as indicated in the following table:

	Uncorrected Sum C_0 8421		Corrected Sum C_0 8421
0	0000		0000
1	0001		0001
2	0010		0010
3	0011	No	0011
4	0100	correction	0100
5	0101	necessary	0101
6	0110		0110
7	0111		0111
8	1000		1000
9	1001		1001
10	1010		1 0000
11	1011		1 0001
12	1100		1 0010
13	1101		1 0011
14	1110		1 0100
15	1111		1 0101
16	1 0000		1 0110
17	1 0001		1 0111
18	1 0010		1 1000
19	1 0011		1 1001

Analysis of the table shows that the correction should be made when the uncorrected sum contains an 8 and 2 or an 8 and 4 or when there is a carry-out from the 8's position. Analysis of the table also shows that the corrected sum can be obtained by adding 0110 (6). This is numerically equivalent to subtracting 1010 (10) and adding 1 0000 (16) (generating a carry-out to the next higher-order position).

The circuit Fig. 6-10 illustrates one decimal position of a BCD adder.

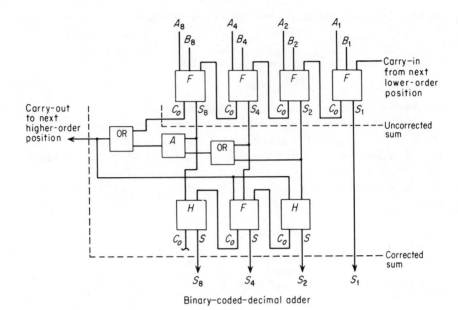

Binary–coded–decimal adder

Figure 6-10

PROBLEMS

1. Convert 111011 (base 2) to base 3.

2. Convert 2601 (base 7) to base 6.

*3. Convert 3333 (base 6) to base 7.

4. Convert 13.8125 (base 10) to base 2.

*5. Convert 49.296875 (base 10) to base 4.

7

Codes, Error Detection, Error Correction

In this chapter, some of the codes used for data representation will be discussed. These codes are used for such things as arithmetic processes and storage and transmission of information. For example, instead of a decimal 6 being represented by a signal on one of ten lines (Fig. 7-1) or by one of ten timed signals (Fig. 7-2) the 6 may be binary-coded as 0110, and represented with only four lines (Fig. 7-3) or with only four timed signals (Fig. 7-4).

The concepts of error detection and correction will also be studied.

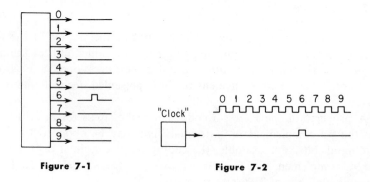

Figure 7-1

Figure 7-2

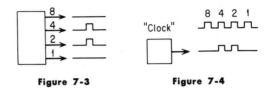

Figure 7-3   Figure 7-4

Nonchecking Numeric Codes

First, various schemes for coding the ten decimal digits, 0 through 9, will be examined.

BCD Code

One of the most logical codes for representing the decimal digits is the binary-coded-decimal or BCD code. Four "bits" (binary digits) are required to code the ten decimal digits.

$$
\begin{array}{c|cccc}
 & 8 & 4 & 2 & 1 \\
\hline
0 & 0 & 0 & 0 & 0 \\
1 & 0 & 0 & 0 & 1 \\
2 & 0 & 0 & 1 & 0 \\
3 & 0 & 0 & 1 & 1 \\
4 & 0 & 1 & 0 & 0 \\
5 & 0 & 1 & 0 & 1 \\
6 & 0 & 1 & 1 & 0 \\
7 & 0 & 1 & 1 & 1 \\
8 & 1 & 0 & 0 & 0 \\
9 & 1 & 0 & 0 & 1 \\
\end{array}
$$

BCD code

Excess-3 Code

The BCD code may be thought of as utilizing the first ten of the sixteen possible combinations of four bits. Another code, which utilizes the middle ten of these sixteen combinations, is called the Excess-3 code. Each coded character is the binary equivalent of the represented decimal number *plus three*.

A property of the Excess-3 code that makes it useful in arithmetic is that the 9's complement of a decimal digit may be obtained by complementing all bits. For example, the coding for the decimal digit 1 is 0100. The 9's complement of 1 is 8, which, in the Excess-3 code, is 1011. Complementing all bits of 0100 results in 1011.

	8	4	2	1
	0	0	0	0
	0	0	0	1
	0	0	1	0
0	0	0	1	1
1	0	1	0	0
2	0	1	0	1
3	0	1	1	0
4	0	1	1	1
5	1	0	0	0
6	1	0	0	1
7	1	0	1	0
8	1	0	1	1
9	1	1	0	0
	1	1	0	1
	1	1	1	0
	1	1	1	1

Excess-3 code { 0 1 2 3 4 5 6 7 8 9

Cyclic Codes

Sometimes it is desirable to have a code in which successive coded characters differ in only one bit position. Such codes are called cyclic codes, and they are particularly useful in analog-digital systems.

One type of cyclic code is the reflected code. A reflected binary code for sixteen decimal digits follows. This code is also known as the Gray code.

Note that except for the high-order position, all columns are "reflected" about the midpoint; in the high-order position, the top half is all 0's and the bottom half all 1's. This pattern can be used for a reflected binary code of any number of bits. A reflected code for three bit positions is enclosed by dotted lines for illustration.

0	0	0	0
0	0	0	1
0	0	1	1
0	0	1	0
0	1	1	0
0	1	1	1
0	1	0	1
0	1	0	0
1	1	0	0
1	1	0	1
1	1	1	1
1	1	1	0
1	0	1	0
1	0	1	1
1	0	0	1
1	0	0	0

Reflected binary or Gray code

If a reflected BCD code is desired, the first ten of the sixteen combinations could be utilized. By choosing the middle ten combinations, instead of the first ten, a reflected Excess-3 code is obtained. An advantage of the reflected Excess-3 code is that the 9's complement can be obtained merely by complementing only the high-order bit, which is an ultimate in ease of complementation. Both reflected codes are shown below:

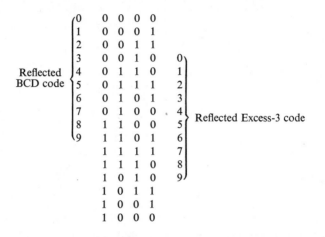

Error Detection and Correction

None of the codes discussed so far can be error-checked. If bits become erroneously changed, say because of circuit failure, there would be no general way to detect the error because in these codes there are cases of two coded characters differing in only one bit position. If even only a single bit in a character became erroneously changed, another valid character could result, and there would be no way of knowing that the resultant character was not the intended one. The characteristics of error detecting and error correcting codes will now be discussed.

The *distance* between two coded characters is the number of bits that must change in one character so that the other character results. For example, the distance between the coded characters 0011 and 1000 is three, since three bits must change to transform one of the characters to the other.

The *minimum distance* of a code is the minimum number of bits that must change in a coded character so that another valid character of the code will result.

In all of the codes discussed so far, the minimum distance was 1: There was at least one case in each code in which a coded character could be changed to another by changing only one bit.

The relationship between the minimum distance of a code and the amount of error detection or correction possible is as follows:

$$M - 1 = C + D, \quad \text{where} \quad C \leqslant D$$

M = minimum distance of a code
C = number of bits in error that can be corrected
D = number of bits in error that can be detected

Since no error can be corrected without being detected, C cannot be greater than D. All possible values for C and D for values of M up to six are tabulated below.

M	C	D
1	0	0
2	0	1
3	0	2
	1	1
4	0	3
	1	2
5	0	4
	1	3
	2	2
6	0	5
	1	4
	2	3

An error detection code is defined according to one less than the minimum error it will not *always* detect. Thus, if a code detects *all* single, double and triple errors, and *some* or *no* quardruple errors, it is called a triple error detecting code. This would still be so even if the code detected *all* quintuple errors. An error correction code is defined in the same manner, that is, according to one less than the minimum error it will not *always* correct.

The relationship between the minimum distance of a code and the amount of error correction or detection possible may be more graphically pictured if a "table-lookup" error detection system is considered. All the valid characters in the code are stored in the table. Each coded character to be checked is compared with the characters in the table. If a character in the table is found to match exactly, it is assumed that no error has occurred; if no character matches exactly, an error has been detected. Whether or not the error can be corrected depends on the minimum distance of the code, as will be seen.

In codes with a minimum distance of one, where two valid characters may differ in only one bit position, a single error in a character could make that character appear like another valid character in the code. This other valid character would be found in the table, and it would be falsely assumed that no error had occurred. Thus, in codes having a minimum distance of one, single errors can fail to be detected. Of course, if single errors can be undetected, multiple errors can be undetected also.

In codes with a minimum distance of two, all coded characters must differ in at least two bit positions. If there is a single bit error in a character, the character cannot possibly match any of those in the table; therefore, all single errors will be detected. Codes with a minimum distance of two are called single error detecting codes. Errors in two or more bits might make the character match exactly some other valid character in the table, and therefore these errors would not be detected.

In codes with a minimum distance of three, all coded characters must differ in at least three bit positions. A character with a single or double bit error cannot match any character in the table; therefore, all single and double errors will be detected. Errors in three or more bits can result in another valid character and therefore these errors cannot be detected.

Minimum distance three codes can be used for single error correction. The key to error correction is that it must be possible to *locate* the bit or bits in error. If a single error occurs in a minimum distance three code, the resulting character will not match exactly any character in the table, but it will come *within one bit* of matching the correct character. It will not come within one bit of matching any other character. To accomplish the correction, the one bit that does not match is changed.

In any code that can be used for correction, correction is "bought" at the expense of detection. If a minimum distance three code is used for correction, and a double error occurs, the resulting character may come within a single bit of matching some other character in the table. Since there is no way of knowing that a double error has occurred, it would be assumed that the single bit was in error, and this bit would be erroneously "corrected." Thus, the error would be compounded, and an incorrect character would result. Minimum distance three codes thus will not detect double errors if they are used to correct single errors.

Summarizing, if minimum distance three codes are used for correction, the location of one bit in error can be determined and the error corrected. Errors in two or more bits can appear to the error correction system as a single error, and an erroneous correction (undetected error) can result. Minimum distance three codes are often referred to as single error correcting codes.

The characters in minimum distance four codes differ in at least four bit positions. Single, double, and triple errors can be detected with these codes, since the resulting character cannot match any of those in the table. Errors in four or more bit positions can result in a character that matches some other valid character in the table, and thus these errors cannot be detected.

Instead of minimum distance four codes being used for triple error detection, they can be used for single error correction with double error detection. If a single error occurs, the resulting character will not match exactly any character in the table, but it will come within one bit of matching the correct character. It will differ from all other characters in the table by at least three bits. The correction is made by changing the one bit that does not match.

A character with a double error will come within two bits of matching the correct character of the table, but it may also come within two bits of matching an incorrect character. There is, therefore, no way of knowing which bits are actually in error and so no attempt is made to correct double errors; they are simply detected.

If a triple error occurs in a character, the resultant character will differ from the correct one in the table in three bit positions, but it may differ from some other character in the table in only one bit position. This one bit would thus be erroneously "corrected," and an incorrect character would result.

Summarizing, if minimum distance four codes are used for correction, the location of one bit in error can be determined and the error corrected. Double errors can be detected but their location cannot be determined for correction. Errors in three or more bit positions can appear to the error correction system as a single error, and an erroneous correction (undetected error) can result,

Minimum distance four codes are often referred to as single error correcting double error detecting codes.

The table-lookup system was used as an aid in learning the concept of minimum distance as it relates to error detection and correction. In practice there are many schemes for accomplishing error detection and correction. As other codes are now examined, it will be seen how some of these schemes work.

Single Error Detection—Minimum Distance Two Codes

A single error detecting code can be obtained by adding a redundant bit to a nonchecking code. The redundant bit can be added to each character in such a way as to make the number of 1 bits in the character even. If this is done, the code is referred to as an "even parity" or "even redundancy" code. The redundant bit may instead be added to each character so as to make the number of 1 bits in the character odd, giving an "odd parity" or "odd redundancy" code. Following are examples of both types of parity codes:

Even Parity BCD Code						Odd Parity BCD Code				
8	4	2	1	R		8	4	2	1	R
0	0	0	0	0	0	0	0	0	0	1
0	0	0	1	1	1	0	0	0	1	0
0	0	1	0	1	2	0	0	1	0	0
0	0	1	1	0	3	0	0	1	1	1
0	1	0	0	1	4	0	1	0	0	0
0	1	0	1	0	5	0	1	0	1	1
0	1	1	0	0	6	0	1	1	0	1
0	1	1	1	1	7	0	1	1	1	0
1	0	0	0	1	8	1	0	0	0	0
1	0	0	1	0	9	1	0	0	1	1

The odd parity *BCD* code is sometimes preferred over the even parity *BCD* code because an all-0 character is frequently undesirable: If a gross circuit failure can change a character to all 0's, it is desirable that an all-0 character not be one of the valid characters in the code. However, a modification is frequently made in which the binary 1010 is assigned to the decimal 0; thus, the even parity character 10100 rather than 00000 represents the decimal 0.

Characters in these codes are checked for the proper parity. If a single error occurs, it will be detected because the character will have the wrong parity. Double errors will not be detected since the parity will check correctly.

Another class of codes are the *fixed-bit* or *m-out-of-n* codes. In these codes there are n bits per character, of which m bits are 1's. Such a code suitable for representing the ten decimal digits is the 2-out-of-5 code, there being exactly ten combinations of five things taken two at a time ($_5C_2$).

While any assignment of the ten 2-out-of-5 combinations to the ten decimal digits could be made, there are some that are more convenient to remember. It is not possible to correctly "weight" all ten combinations, but it is possible to properly weight nine of them. Two such weightings are shown: the 01247 code and the 01236 code:

2-out-of-5 Codes

0	1	2	4	7		0	1	2	3	6
0	0	0	1	1	0	0	1	1	0	0
1	1	0	0	0	1	1	1	0	0	0
1	0	1	0	0	2	1	0	1	0	0
0	1	1	0	0	3	1	0	0	1	0
1	0	0	1	0	4	0	1	0	1	0
0	1	0	1	0	5	0	0	1	1	0
0	0	1	1	0	6	1	0	0	0	1
1	0	0	0	1	7	0	1	0	0	1
0	1	0	0	1	8	0	0	1	0	1
0	0	1	0	1	9	0	0	0	1	1

Only the decimal 0 is improperly weighted in both codes.

There are two other 2-out-of-5 codes that weight nine of the ten combinations correctly, but both of these involve negative weights:

$$-1, 2, 3, 4, 5$$
$$-2, 1, 3, 4, 5$$

A popular 2-out-of-5 code that weights eight of the ten combinations properly is the 84210 code. All combinations are weighted correctly except the decimal 0 and 7; the 8—2 combination is used for the decimal 0, and the 8—4 combination is used for the decimal 7. This code is not very different from the even parity BCD code and very little logic circuitry is needed to convert from one of these codes to the other.

Characters in fixed bit codes are checked for the correct number of 1-bits. Single errors will be detected since the number of 1-bits in the character will be one too many or one too few. Double errors involving two 1's or two 0's will also be detected, but double errors in which a 0 becomes a 1, and a 1 becomes a 0, will not be detected.

Another popular fixed bit code is the biquinary code. This is a seven-bit code made up of a 1-out-of-2 group and a 1-out-of-5 group. Again, there are ten possible combinations. Two possible weightings for this code are shown; the second one is sometimes called the "quibinary code" to differentiate it from the first one:

Biquinary Codes

0 5	0 1 2 3 4		0 1	0 2 4 6 8
1 0	1 0 0 0 0	0	1 0	1 0 0 0 0
1 0	0 1 0 0 0	1	0 1	1 0 0 0 0
1 0	0 0 1 0 0	2	1 0	0 1 0 0 0
1 0	0 0 0 1 0	3	0 1	0 1 0 0 0
1 0	0 0 0 0 1	4	1 0	0 0 1 0 0
0 1	1 0 0 0 0	5	0 1	0 0 1 0 0
0 1	0 1 0 0 0	6	1 0	0 0 0 1 0
0 1	0 0 1 0 0	7	0 1	0 0 0 1 0
0 1	0 0 0 1 0	8	1 0	0 0 0 0 1
0 1	0 0 0 0 1	9	0 1	0 0 0 0 1

An advantage of these codes is that the circuitry to perform arithmetic operations is quite economical. The quibinary code has the advantage of more economical conversion to and from the BCD code.

Single Error Correction—Minimum Distance Three Codes

The construction and operation of a *Hamming code* will be used as an example in the study of single error correcting codes.

First of all, in determining how many bits per character are required, the bit positions are numbered sequentially from left to right as 1, 2, 3, etc. The positions that are a power of two, that is, positions 1, 2, 4, 8, 16, etc., are reserved for check bits. All other bit positions may then contain information bits.

If a single error correcting numeric code is required, and the four-bit BCD code is used for the information, seven bits in all would be required: positions 1, 2, and 4 for check bits, and positions 3, 5, 6, and 7 for the four information bits. These seven bits can be labeled as follows:

$$1 \quad 2 \quad 3 \quad 4 \quad 5 \quad 6 \quad 7$$

$$C_1 \quad C_2 \quad 8 \quad C_4 \quad 4 \quad 2 \quad 1$$

The values of the check bits C_1, C_2, and C_4, for each coded character, are determined as follows:

C_1 is chosen so as to establish even parity for positions 1, 3, 5, and 7.
C_2 is chosen so as to establish even parity for positions 2, 3, 6, and 7.
C_4 is chosen so as to establish even parity for positions 4, 5, 6, and 7.

This pattern may be more obvious if the position locations are written in binary.

(C_1)	1	0	1	0	1	0	1
(C_2)	0	1	1	0	0	1	1
(C_4)	0	0	0	1	1	1	1
	1	2	3	4	5	6	7
	C_1	C_2	8	C_4	4	2	1

For an example, the check bits for the character for the decimal 9 will be generated:

$$1 \quad 2 \quad 3 \quad 4 \quad 5 \quad 6 \quad 7$$

$$C_1 \quad C_2 \quad 8 \quad C_4 \quad 4 \quad 2 \quad 1$$

$$1 \qquad 0 \quad 0 \quad 1$$

C_1 must be chosen so as to establish even parity for positions 1, 3, 5, and 7; therefore, C_1 must be a 0:

1	2	3	4	5	6	7
C_1	C_2	8	C_4	4	2	1
0		**1**		**0**	**0**	**1**

C_2 must be chosen so as to establish even parity for positions 2, 3, 6, and 7; C_2 must also be a 0:

1	2	3	4	5	6	7
C_1	C_2	8	C_4	4	2	1
0	**0**	1		0	0	1

C_4 must be chosen so as to establish even parity for positions 4, 5, 6, and 7; C_4 must therefore be a 1:

1	2	3	4	5	6	7
C_1	C_2	8	C_4	4	2	1
0	0	1	**1**	**0**	**0**	**1**

The coded character for the decimal 9 is therefore

$$0011001$$

To illustrate how a single error in this coded character can be detected and corrected, an error will be "made" in position 6:

1	2	3	4	5	6	7
0	0	1	1	0	1	1

The three parity checks, involving C_1, C_2, and C_4, are applied to the character. Based on the outcome of these checks, a binary number is developed, the C_1, C_2, and C_4 checks corresponding, respectively, to the 1, 2, and 4 positions of the binary number. If the check shows even (correct) parity, a 0 is entered in the corresponding position of the binary number; if the check shows odd (incorrect) parity, a 1 is entered. The resulting binary number indicates the position in error; to correct the error, the bit in the position indicated is changed.

In this example, the three parity checks are as follows:

$$
\begin{array}{llllllllll}
C_1: & \mathbf{0} & 0 & \mathbf{1} & 1 & \mathbf{0} & 1 & \mathbf{1} & \text{even} \\
C_2: & 0 & \mathbf{0} & \mathbf{1} & 1 & 0 & \mathbf{1} & \mathbf{1} & \text{odd} \\
C_4: & 0 & 0 & 1 & \mathbf{1} & \mathbf{0} & \mathbf{1} & \mathbf{1} & \text{odd}
\end{array}
$$

$$
\begin{array}{ccc}
4 & 2 & 1 \\
1 & 1 & 0 & = 6
\end{array}
$$

The C_1 parity check shows even parity, while the C_2 and C_4 parity checks show odd parity. The resulting binary number, $110 = 6$, indicates that position 6 is in error. To correct the error, the bit in position 6 is changed from a 1 to a 0. Study of the construction of this code will show that the position of any bit in error is uniquely identified by the outcome of the parity checks. If the resultant binary number is zero (000), no error is indicated.

Only single errors are detected and corrected with this code. Errors in two bit positions will appear to the error correction system as a single error, and a false correction will be made. Triple errors may also appear as single errors and be falsely corrected, or they may "cancel out" and appear as no error.

If this code is used for detection only, all single and double errors will be detected; the error detection system checks only for the occurrence of an error, but does not try to identify a position in error and correct it.

This Hamming code is by no means the only one allowing single-error correction. As long as the coded characters are chosen so that all pairs of characters differ in *at least* three bit positions, single-error correction can be accomplished.

For a given number of check bits, C, the maximum number of information bits, I_{max}, is given by

$$
I_{max} = 2^C - C - 1
$$

Single Error Correction with Double Error Detection—Minimum Distance Four Codes

Hamming single error correcting codes can be extended into Hamming single error correcting double error detecting codes simply by the addition of one more bit establishing even parity over the entire coded character.

For example, taking the seven-bit character for the decimal 9, if an eighth bit is added to establish even parity over the entire character, this bit must be a 1, and the resulting character for the 9 is

00110011

Four parity checks are made on the character: the C_1, C_2, and C_4 checks, and the overall parity check, which can be called the P check. Any single error will be indicated by the P check showing odd parity. If the single error occurs in any of the first seven-bit positions, it will show up in some combination of the C_1, C_2, and C_4 checks, which will indicate the position in error. If the single error is in the eighth bit, the absence of any error indication by the C_1, C_2, and C_4 checks indicates that the bit in error is the eighth bit. The P check showing odd parity thus indicates that a single error has occurred, and that a correction should be made.

If a double error occurs, the P check will show even parity. Even though the C_1, C_2, and C_4 checks indicate some position in error, the P check showing even parity indicates that a double error has occurred, and that no correction should be made.

Alphanumeric Codes

"Alphanumeric" codes are those containing enough coded characters to code the ten decimal digits, the twenty-six letters of the alphabet, and often special symbols also. A few examples of such codes will be described.

The BCD code can be expanded into a six-bit code giving 64 possible coded characters, satisfying the requirement for an alphanumeric code. To make such a code into a single-error-detecting code, a parity bit can be added.

There are various *m-out-of-n* codes used for alphanumeric information. For instance, 3-out-of-8 codes (56 characters) and 4-out-of-8 codes (70 characters) are used.

Alphanumeric Hamming single-error-correcting codes require ten bits in all, six information bits and four check bits, the check bits occupying positions 1, 2, 4, and 8. By adding an eleventh bit to this code, an alphanumeric Hamming single error correcting double error detecting code is obtained.

Cross-Parity

Sometimes a check is associated with an entire block of characters. For instance, at the end of a block of even parity BCD characters, an entire redundant character is added, the bits in this character being chosen so as to establish even parity in each "channel," that is, the 8-bit channel, the 4-bit channel, the 2-bit channel, etc.

Parity checks on this block of information are made in two directions: "vertically" for each character and "horizontally" for each channel. This code will detect all single, double and triple errors, or it may be used as a single error correcting double error detecting code.

EXAMPLE

Characters

	9	5	7	1	4	3	8	2	R
8	1	0	0	0	0	0	1	0	0
4	0	1	1	0	1	0	0	0	1
2	0	0	1	0	0	1	0	1	1
1	1	1	1	1	0	1	0	0	1
R	0	0	1	1	1	0	1	1	1

Suppose that an error occurs in the "1" bit of the "3" character.

Characters

	9	5	7	1	4	3	8	2	R	
8	1	0	0	0	0	0	1	0	0	
4	0	1	1	0	1	0	0	0	1	
2	0	0	1	0	0	1	0	1	1	
1	1	1	1	1	0	**0**	0	0	1	<
R	0	0	1	1	1	0	1	1	1	

$\wedge$

The odd vertical parity on the "3" character and the odd horizontal parity on the "1" channel locate the single error for correction.

PROBLEM

1. Each digit of a 4-digit decimal number is encoded in a single error correcting double error detecting Hamming code. The coded information is received high-order digit first as follows:

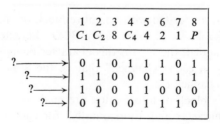

	1	2	3	4	5	6	7	8
	C_1	C_2	8	C_4	4	2	1	P
?⟶	0	1	0	1	1	1	0	1
?⟶	1	1	0	0	0	1	1	1
?⟶	1	0	0	1	1	0	0	0
?⟶	0	1	0	0	1	1	1	0

Correct and decode:

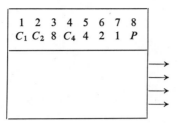

8

Introduction to
Sequential Circuits

The circuits that have been discussed so far are called *combinational
circuits*. In combinational circuits, the output states are functions solely of
the present input states: For a particular input combination there will either
always be an output or else there will never be an output.

The rest of the chapters in this book discuss *sequential circuits*. In sequential circuits, the output states are functions not only of the present input
states but also of the present *circuit state*. Circuit states, in turn, are functions
of past circuit states, which, in turn, are functions of past input states. Sequential circuits thus have a memory, the output states being functions not only
of the present input states but also of past input states.

An example of a simple sequential circuit requirement is as follows.
A circuit is to have two inputs, x_1 and x_2, and one output, Z. The output is
to be *on* only when the inputs are in the state $x_1 x_2 = 10$, immediately following the state $x_1 x_2 = 11$. Note a characteristic of a sequential circuit:
when the inputs are in the state $x_1 x_2 = 10$, the output should sometimes be
other times be *off*, depending on what has happened before. For the
"remember" information about past circuit states, some sort of
required.

matic diagram of a generalized sequential circuit is shown in
inputs to a sequential circuit, sometimes referred to as *primary*

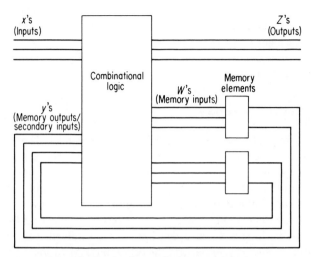

Schematic diagram of generalized sequential circuit

Figure 8-1

inputs, are conventionally represented by x's. The outputs of a sequential circuit are represented by Z's. The memory element inputs are here represented by W's, for generality. There are many types of memory elements, and later on, as the different types are examined, other symbols characteristic of each type will be substituted for the W's in each specific case.

The memory element outputs reflect the states of the memory elements, and are represented by y's. The memory outputs serve as *secondary inputs* to the circuit.

The memory elements may have one or more inputs and either a single output, y, or two complementary outputs, y and $\bar{y}$. As in combinational logic blocks, the W's and y's assume the values 0 or 1. The states of the W's are called the memory *excitations,* and the resulting states of the y's are called the memory *responses.*

A property of the memory elements is that there is a time delay between a change in their excitation and the resulting response, that is, between the time there is a change in the W's and a change occurs in the y's. This time is referred to as the *transition time* of the memory element. Another property of the memory elements is that they can remain in a state, y, after the excitation that caused that state is no longer present.

The Z and W states are, in general, functions of the x and y states. Consider, however, only the W's and y's for the moment. A "present" W state is a function of the "present" y states (potentially, all of the y's), this function being implemented by the combinational logic. A "next" y state is a function of the "present" W states for each memory element, this function being dependent on the type of memory element involved. These functions, in fact, are what characterize each type of memory element; the functions of several types will be described in a later chapter. The "next" y state, after the transition time delay of the memory element, will be realized and become the new "present" y state.

Pulse and Level Operation

There are several types of sequential circuit operation. These types have been classified in various ways, under such terms as pulse, level, clocked, synchronous, asynchronous, and fundamental.

In the synthesis of sequential circuits, there are certain steps that are followed. Some of these steps are independent of the type of operation, while others are dependent on the type. In terms of affecting the synthesis procedure, the various types of operation can be resolved into two classes; one we shall call *pulse operation*, and the other, *level operation*.

The distinction between pulse and level operation involves the duration of the memory element excitation in relationship to the memory element transition time. More specifically, in pulse operation, the memory excitation is gone before the memory element responds; in level operation, the memory excitation is still present when the memory element responds. These timing relationships are shown in Fig. 8-2. Note that it is the timing of the memory element inputs, not the circuit inputs, that defines the type of operation.

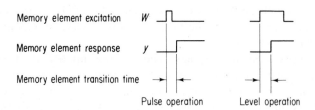

Memory element excitation W

Memory element response y

Memory element transition time

Pulse operation   Level operation

Figure 8-2

Signals are described as being *pulses* or *levels*. A pulse is somewhat vaguely defined as a relatively short-duration signal; a level is similarly vaguely defined as a relatively long-duration signal. The important distinction between pulses and levels, as it affects the synthesis procedure, is that just discussed in connection with the timing of the memory element excitation.

With reference to the general sequential circuit model (Fig. 8-1), the x signals may be pulses or levels; the y signals, from the memory element outputs, are always levels.[1] Since the W's and Z's are, in general, functions of the x's and y's, the W and Z signals may be pulses or levels; however, for a W or Z signal to be a pulse, at least one x signal must be a pulse. Note that the y signals are levels whether the W signals are pulses or levels.[1]

In pulse operation, the W signals are pulses; therefore, at least one x signal must be a pulse; the Z signals, however, may be pulses or levels (depending on whether the Z's are functions of the x's and y's, or only of the y's). In level operation, all signals must be levels.

Memory elements that respond to pulses are described as being pulse-sensitive; memory elements that respond to levels are said to be level-sensitive. Some memory elements are described as edge-sensitive, since they respond to the positive-going or negative-going edge of a signal (see Fig. 8-3). In sequential circuit synthesis, edge-sensitive design can be treated in the class of pulse operation, since the excitation is gone before the memory element responds.

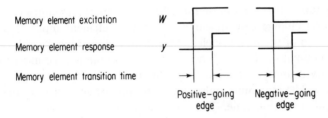

Edge–sensitive memory element response

Figure 8-3

Assuming that we start with a stable sequential circuit, the following sequences of events are possible. Taking pulse operation first, an x pulse occurs. Z or W pulses may (or may not) occur; if no W pulse occurs, circuit action ends. If any W pulses occur, the related memory elements may change stable state, with an accompanying change in the corresponding y levels. If no y level changes, circuit action ends. If any y levels change, Z levels may change. Circuit action ends.

In level operation, circuit action starts with x level changes. Z or W levels may change; if no W level changes, circuit action ends. If any W levels change, the related memory elements may change state, with an accompanying change in the corresponding y levels. If no y level changes, circuit action ends. If any y levels change, Z or W levels may change, and the preceding cycle repeats. Circuit action ceases when no further W or y level changes.

---

[1] There is one memory element for which there is a theoretical exception to this, but from a practical standpoint the statement is true. The theoretical exception will be discussed later.

Note that in pulse operation, the W's are not immediately affected by the new y states; rather the new y states set the stage for the next x pulse to affect the W's. In level operation, however, the W's can be immediately affected by the new y states, and circuit action can continue. Note also that since a W can be a function of its related y, a memory element can enter into its own control.

Restrictions on Operation

Some restrictions on pulse and level operation will now be examined. In pulse operation, the following requirements must be satisfied:

1. Minimum pulse: The pulse must be long enough to cause the memory element to change state.
2. Maximum pulse: The pulse must be gone before a memory element output level change can appear at a memory element input. This time is composed of the memory element transition time plus the delays through the combinational logic. Since most of this time is accounted for by the memory element transition time, this requirement is generally stated as follows: The maximum pulse duration must be smaller than the memory element transition time. The purpose of this requirement is to prevent further changes in memory elements (until the next circuit input pulse).
3. Minimum time between the start of successive pulses: This time must be longer than the memory element transition time plus resolution time (the time that the memory element needs to stabilize internally before it is ready to accept another pulse) and the delays through the combinational logic (see footnote 1). In other words, the entire circuit must be stable before another pulse occurs. This requirement applies to successive pulses whether on the same or different input lines.

With more than one memory element in a circuit, safest or "worst case" design would consider the memory element with the shortest response time for requirement 2, and the longest response time plus resolution time for requirement 3.

The restriction in requirement 3 prohibits "simultaneous" input pulses. Theoretically, simultaneous input pulses could be allowed if they could be guaranteed to arrive precisely together; practically, this cannot be achieved, and reliable operation must forbid pulses on more than one input line at a time.

This restriction applies to simultaneous input pulses but not necessarily to simultaneous input signals; if some of the input signals are pulses and some are levels, the input pulses can occur only one at a time, but the input levels may occur in all possible combinations. Each input level combination must be ANDed with an input pulse. Each input pulse may be ANDed with any or all input levels, or may be used alone. Since all input level combinations are mutually exclusive, and input pulses are required to be mutually exclusive from a timing standpoint, the important consequence is that only one pulse at a time can reach the memory elements.

The input levels must not change while the input pulses are present. This timing relationship is shown in Fig. 8-4.

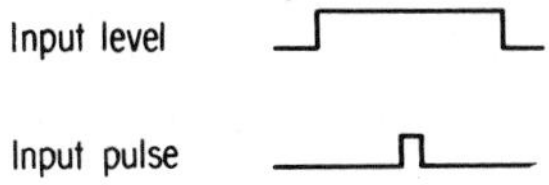

Input pulses that are ANDed with input levels are commonly called *clock pulses*, and the corresponding inputs themselves are called *clocks*. Generally, the level inputs are the "logic" inputs,

Input level

Input pulse

Figure 8-4

while the clocks are used to govern the time that the level inputs are sampled. Some logic designers, in fact, prefer not to think of the clock as an input at all, but rather as a separate entity. For distinguishability, the level inputs are denoted by x's and the clocks by c's. Usually a clock produces a sequence or train of pulses that is evenly timed, or periodic, but this is not a requirement.

In level operation, the following requirement must be satisfied:

Minimum time between changes in input level: An input level change may cause changes in memory element states; no further change in input level can occur until all circuit action ceases, that is, until the circuit is totally stable. This requirement applies to successive level changes whether on the same or different input lines. The "worst case" situation, or longest circuit action time, should govern this requirement.

This restriction prohibits requiring or depending on simultaneous input level changes. The reasoning is analogous to that for pulse operation: simultaneous input level changes cannot be guaranteed to occur exactly together in practical design.

We may, however, allow for the *possibility* of simultaneous input level changes in the event that one or more of the input variables are random in character. There are thus two possibilities in design: (1) If simultaneous input level changes may occur, such possible changes can be taken into account; or (2) if the inputs are controlled so that only single input level changes are allowed to occur, simultaneous input changes are treated as optional combinations in the design.

Other Nomenclature

Most terms relating to sequential circuit design have been misused to the extent that there is hardly a term that remains unambiguous. Some other terms used in the literature follow.

What we here call pulse operation without clocks is also called *pulse mode*. Pulse operation with clocks is also called *clocked* or *synchronous* operation. Level operation is also called *fundamental mode* (simultaneous input changes not allowed) or *asynchronous* operation.

The term "synchronous" actually means having the same period, and implies a regular clock; in usage, however, regular periods are not necessarily implied. Note also that "synchronous" and "asynchronous," as just used, are not opposites. The reader is cautioned that many of these terms also have other meanings in the literature.

Intuitive Look at Sequential Circuit Synthesis

An intuitive look at sequential circuit synthesis will help give a better appreciation of what the formal method of synthesis accomplishes. Some type of sequence of input (x) states is given, and the required sequence of associated output (Z) states is specified. For some input states that appear more than once, the associated output states will not always be the same. When multiple occurrences of the same input state require different associated output states, these occurrences must be differentiated by secondary input (y) states.

These secondary inputs are memory element outputs; therefore, some number of memory elements is required. To obtain the needed memory element output states, the proper memory element input (W) states, or excitations, must be generated. Sometimes, to be able to accomplish this, multiple occurrences of the same input state for which the associated output states are the *same* may also have to be differentiated by secondary input states. Any additional secondary input states may require additional memory elements and the generation of additional excitations. Finally, the circuit output (Z) states are obtained.

In the process of choosing the actual secondary input/memory element output (y) states to be used, many variations are generally possible, and the choice can affect circuit economy. Each variation in the choice of memory element states can result in different excitation (W) and circuit output (Z) expressions and associated combinational logic. It is generally economical, however, to keep the number of memory element states and, even more importantly, the number of memory elements to a minimum.

The formal method of sequential circuit synthesis accomplishes what has just been discussed, but does so, for the most part, in a different and systematic manner. The formal method also considers other problems in the design that are not mentioned here. An overview of the steps in the formal synthesis procedure follows.

Steps in Sequential Circuit Synthesis

1. State diagram, state table, flow table: From the word statement of the problem, a state table (and optionally, a state diagram) is constructed for pulse operation, or a flow table is constructed for level operation.
2. Table reduction: Any redundancy in the table is eliminated, and the table is reduced to the minimum number of rows.
3. State assignment: A combination of memory element (y) states is assigned to each state in the table.
4. Excitation maps and expressions: Memory excitation maps are obtained from the table, and the excitation (W) expressions are read from the maps.
5. Circuit output maps and expressions: Output maps are obtained from the table, and the output (Z) expressions are read from the maps.
6. The sequential circuit is implemented from the excitation and output expressions.

9

State Diagrams, State Tables, Flow Tables

The terms "state" and "flow" as applied to sequential circuit diagrams and tables have been used ambiguously and sometimes interchangeably in the literature. We shall here relate *state diagrams* and *state tables* to pulse operation, and *flow tables* to level operation.

The synthesis of a sequential circuit starts with transforming the word statement of the problem into a state table, in the case of pulse operation, or a flow table, in the case of level operation. In pulse operation, a state diagram may also be drawn; the diagram contains the same information as the table, but permits a more graphic visualization of the entire circuit operation. It is sometimes helpful for the logic designer to construct the state diagram before he constructs the state table.

Maps are eventually obtained from the tables; these maps are read to give the memory element excitation and circuit output expressions.

Before discussion of how the diagrams and tables are constructed from the word statement, the structure of the diagrams and tables themselves will be examined.

State Diagram and State Table

Examples of state diagrams and state tables for pulse operation are shown in Figs. 9-1 and 9-2. In a state diagram, each circuit state is represented by an

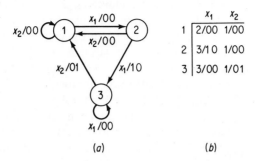

(a)   (b)

Figure 9-1

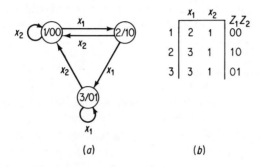

(a)   (b)

Figure 9-2

arbitrary number which is circled. All circuit states are stable, and transitions from state to state are effected by input pulses. Each of these transitions is represented on the diagram by an arrow labeled with the input pulse causing the transition; the arrow leaves the circled number representing the "present" state, and terminates at the circled number representing the "next" state. Although not illustrated in these examples, more than one transition may share the same arrow.

The next state may be the same as the present state; for example, in Figs. 9-1(a) and 9-2(a), if the circuit is in state 1 and an x_2 pulse occurs, the circuit remains in state 1; if the circuit is in state 3 and an x_1 pulse occurs, the circuit remains in state 3.

Pulse outputs are associated with transitions and the corresponding input pulses. On the state diagram, output pulses are labeled adjacent to the associated transition input pulse labels; the entire label takes the format, "$x_n/Z_1Z_2 \ldots Z_n$ values." In Fig. 9-1(a), for example, a Z_1 output pulse is coincident with an x_1 pulse ($x_1/10$) occurring when the circuit is in state 2, and a Z_2 output pulse is coincident with an x_2 pulse ($x_2/01$) occurring when the circuit is in state 3.

Level outputs are associated with circuit states. On the state diagram, output levels are labeled adjacent to the associated circuit state numbers; the entire label takes the format, "state number/$Z_1Z_2 \ldots Z_n$ values." In Fig. 9-2(a), for example, the Z_1 level output is on when the circuit is in state 2 (2/10), and the Z_2 level output is on when the circuit is in state 3 (3/01).

A state table contains the same information as a state diagram. There is a row in the table for each circuit state, and a column for each pulse input. Each circuit state is assigned an arbitrary number which labels the corresponding row; each column is labeled with a pulse input.

Each row label represents a present state. Each entry in the table indicates the next state that will be reached when the circuit is in the present state labeling that row, and the input pulse labeling that column occurs.

Output pulses are labeled adjacent to the entries corresponding to the associated transitions. Output levels are designated at the right of the rows labeled by the associated circuit states. The reader should verify that the state tables in Figs. 9-1(b) and 9-2(b) contain the same information as each corresponding state diagram.

Clocked Pulse Operation

In clocked pulse operation, transitions from state to state are effected by clock pulses ANDed with level input combinations, or by clock pulses alone. Some memory elements, as a part of their function, can perform this ANDing internally. Each transition arrow on the state diagram is labled accordingly, the general format being "$c_n/x_1x_2 \ldots x_n$." With pulse outputs, the entire transition label takes the general format "$c_n/x_1x_2 \ldots x_n/Z_1Z_2 \ldots Z_n$ values." If there is a single clock (a most common situation), it is usually not explicitly indicated but is understood.

In general, each clock may be used in conjunction with all or just some of the level inputs, or may be used alone. If n is the number of level inputs associated with a particular clock, the number of distinct pulses from this clock is 2^n. The total number of distinct pulses in the circuit is obtained by summing these values of 2^n for all clocks. The maximum number of distinct pulses possible, which occurs if every clock is used in conjunction with every level input, is $m \cdot 2^p$, where m is the total number of clocks and p is the total number of level inputs.

All combinations of clock pulse and level inputs, or clock pulses alone, are mutually exclusive; they could be thought of as external to the circuit, and the combinations renamed simply $x_1, x_2, \ldots, x_n$. In this form, the state diagram and table would appear exactly as in pulse operation, accentuating the fact that clocked pulse operation is a special case of pulse operation.

Examples of state diagrams and tables for clocked pulse operation are shown in Figs. 9-3 (pulse output) and 9-4 (level output). A single clock is implied in both examples.

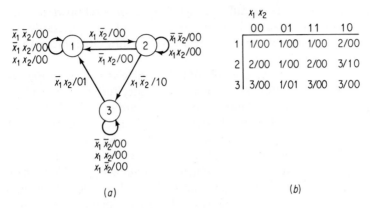

	$x_1 x_2$			
	00	01	11	10
1	1/00	1/00	1/00	2/00
2	2/00	1/00	2/00	3/10
3	3/00	1/01	3/00	3/00

(a)

(b)

Figure 9-3

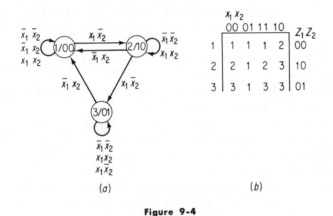

	$x_1 x_2$				$Z_1 Z_2$
	00	01	11	10	
1	1	1	1	2	00
2	2	1	2	3	10
3	3	1	3	3	01

(a)

(b)

Figure 9-4

Flow Table

In pulse operation, all states are stable, and transitions (which are due to input pulses) are always from stable state to stable state. In level operation, there are both stable and unstable states, and transitions (which are due to input level changes) may lead from a stable state through one or more unstable states before the next stable state is reached.

In a table for level operation, an entry may represent either a stable or unstable state. If the entry represents an unstable state, the entry number indicates what the *next stable state* will be. The entry does not, however, always indicate what the *next state* will be; the next state may be another unstable state with the same entry number.

Since the entries have a different significance for level operation than for pulse operation, indicating the next stable state but not necessarily the next state, the tables for level operation are here called flow tables rather than state tables, for differentiation.

An example of a flow table is shown in Fig. 9-5. In a flow table, there is an entry for every possible circuit state. Some of these states are stable and some are unstable. Stable states are indicated by circled numbered entries, and unstable states by uncircled numbered entries. An uncircled number corresponds to the next stable state in which the circuit action will terminate.

$x_1 x_2$

00	01	11	10	$z_1 z_2$
①	2	–	3	00
1	②	4	3	00
1	2	5	③	00
1	2	④	3	01
1	2	⑤	3	10

Figure 9-5

For example, in Fig. 9-5, the circled ⑤ entry, which is associated with the input combination $x_1 x_2 = 11$, represents a stable state. If the circuit is in stable state ⑤, and the input combination, for example, is changed to $x_1 x_2 = 01$, the circuit enters unstable state 2. The 2 entry indicates that the next stable state will be ②; there is no information at this time as to what the next state will be.[1]

The output states in each row are associated with the circled stable state entry in that row.

The reflected ordering is used for the input combinations since the table will eventually be transformed into a map.

While only the *next stable state* is specified for each entry during the construction of the flow table from the word statement, later modifications to the table will eventually specify the *next state* in each case, and the flow table will be transformed into a state table.

A pulse input/pulse output circuit, such as represented by Figs. 9-1 and 9-3, is called a *Mealy model*. A pulse input/level output circuit, such as represented by Figs. 9-2 and 9-4, is called a *Moore model*. A level input/level output circuit, such as represented by Fig. 9-5, is called a *Huffman model*.

Word Statement to State Diagram, State Table, and Flow Table

The general approach to obtaining the diagram or table from the word statement of the problem will now be discussed. The wording of the next

---

[1]We could regard pulse operation state tables in a somewhat comparable way. The column of row designations comprises all of the stable states; these row designations correspond to the flow table circled entries. This column represents a "no pulse," or "no input," column.

No stable state is associated with an input; each entry in the table denotes the next stable state that will be reached. These entries correspond to the flow table uncircled entries.

three paragraphs applies directly to state diagrams and tables; for flow table, substitute "stable state" for "state."

Start with an initial state, 1, and add states as required. For each combination of present state and input, determine the next state. The next state may be the same as the present state, it may be a different state that already exists, or a new state may have to be added. Continue this process until all possible transitions from each existing state have been considered; the process will then be complete.

In the interests of economy, transitions to existing states should be considered first, and a new state added only if necessary. However, if redundant states are included, there are methods for removing them (Chapter 10).

The number of states needed may sometimes be determined directly from the problem statement before the start of construction, but more commonly it is determined during the construction process.

Word Statement to State Diagram and State Table

The construction of the state diagram and state table from the word statement of the problem will now be illustrated by the use of some examples involving pulse operation. The circuit requirements in the first four examples involve two pulse inputs, x_1 and x_2, and one pulse output, Z.

EXAMPLE § 1

An output pulse Z is to be coincident with the first x_2 pulse immediately following an x_1 pulse.

Refer to Fig. 9-6. Start with initial state, 1. The circuit will remain in state 1 as long as x_2 pulses occur; an x_1 pulse will take the circuit to a new state, 2. The circuit will remain in state 2 for any further x_1 pulses. The first x_2 pulse occurring while the circuit is in state 2 will be the first immediately following an x_1 pulse; this pulse will produce the desired output pulse and, at the same time, will return the circuit to state 1.

It is assumed that circuit state 1 is always the "power on" state, that is, when the power is first turned on, the circuit will be in state 1. If the problem statement were slightly revised to read "An output pulse Z is to be coincident with the first of a sequence of consecutive x_2 pulses," the state diagram and table in Fig. 9-7 would result. A difference between the two definitions occurs

	x_1	x_2
1	2/0	1/0
2	2/0	1/1

Figure 9-6

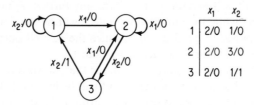

	x_1	x_2
1	1/0	2/1
2	1/0	2/0

Figure 9-7

only if the very first pulse to occur is an x_2 pulse. In the first case, no output pulse will occur, since the x_2 pulse is not preceded by an x_1 pulse; in the second case, an output pulse will occur.

Once the circuit is in operation, however, the circuit actions in both cases are identical. Examination will show that Figs. 9-6 and 9-7 are identical except for the interchange of states 1 and 2. Consideration of initial sequences can therefore be of importance.

It is suggested that, for practice, the reader draw his own diagrams and tables for the examples that follow, and compare his results with those shown.

EXAMPLE § 2

An output pulse Z is to be coincident with the second of a sequence of consecutive x_2 pulses immediately following an x_1 pulse.

Refer to Fig. 9-8. The circuit will remain in state 1 as long as x_2 pulses occur; an x_1 pulse will take the circuit to a new state, 2. The circuit will remain in state 2 for any further x_1 pulses; an x_2 pulse, the first immediately following an x_1 pulse, will take the circuit to a new state, 3. If in state 3 an x_1 pulse occurs, the circuit will be taken back to state 2; however, if in state 3 an x_2 pulse occurs, it will be the second consecutive x_2 pulse immediately following an x_1 pulse; this pulse will produce the desired output pulse and return the circuit to state 1.

	x_1	x_2
1	2/0	1/0
2	2/0	3/0
3	2/0	1/1

Figure 9-8

The rest of the examples are given without comment.

EXAMPLE § 3

An output pulse Z is to be coincident with the first x_2 pulse immediately following two or more consecutive x_1 pulses. (See Fig. 9-9.)

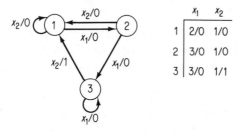

Figure 9-9

EXAMPLE § 4

An output pulse Z is to be coincident with the first x_2 pulse immediately following exactly two consecutive x_1 pulses. (See Fig. 9-10.)

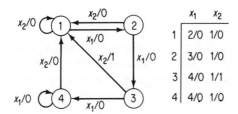

Figure 9-10

The circuit requirements in the next two examples involve two pulse inputs, x_1 and x_2, and one level output, Z.

EXAMPLE § 5

The output is to turn on with an x_2 pulse. The output is to turn off with the second of a sequence of consecutive x_1 pulses immediately following an x_2 pulse. No other input sequence is to cause any change in output. (See Fig. 9-11.)

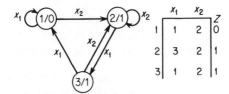

Figure 9-11

s to turn on with the first of a sequence of x_2 pulses. The out-
ff with the second of a sequence of consecutive x_2 pulses. No
ience is to cause any change in output. (See Fig. 9-12.)

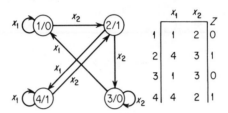

	x_1	x_2	Z
1	1	2	0
2	4	3	1
3	1	3	0
4	4	2	1

Figure 9-12

The next two examples involve a clock.

EXAMPLE § 7

A sequential circuit is to have two level inputs, x_1 and x_2, and one clock.
An output pulse, Z_1 is to be coincident with a clock pulse occurring with
$x_1x_2 = 01$ immediately following two or more consecutive clock pulses with
$x_1x_2 = 10$. $x_1x_2 = 00$ and $x_1x_2 = 11$ can never occur. (See Fig. 9-13.)

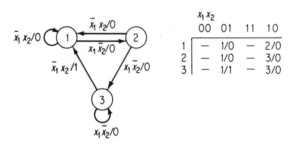

	$x_1 x_2$			
	00	01	11	10
1	—	1/0	—	2/0
2	—	1/0	—	3/0
3	—	1/1	—	3/0

Figure 9-13

The reader should compare this example with Example § 3.

EXAMPLE § 8

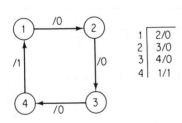

1	2/0
2	3/0
3	4/0
4	1/1

Figure 9-14

A sequential circuit is to have one clock.
An output pulse, Z_1 is to be coincident with
every fourth clock pulse. (See Fig. 9-14.)

Note that in clock pulse operation in
which the only inputs are clocks (no level
inputs), the next state is solely a function
of the present state.

Word Statement to Flow Table

In general, a row of a flow table may contain none, one, or more than one stable state. A flow table in which each stable state is assigned a separate row is called a *primitive flow table*. To obtain a flow table with the minimum number of rows, one must, in general, start with a primitive flow table; the concept is somewhat analogous to that in the tabular method of minimization, in which one starts with the expanded form. A primitive flow table is therefore constructed from the word statement of the problem. An example follows.

EXAMPLE

A sequential switching circuit is to have two level inputs, x_1 and x_2, and one level output, Z. Z is to turn on when x_2 turns on, provided x_1 is already on. Z is to turn off when x_2 turns off. No other input change is to cause any change in output. Only one input can change state at a time.

Development of the primitive flow table can be started by first considering the sequence for turning on the output (Fig. 9-15). All optional entries are due to the restriction to single changes of input.

If the circuit is in stable state ②, and the inputs change from $x_1 x_2 = 10$ to $x_1 x_2 = 00$, the circuit can return to stable state ①

$x_1 x_2$				Z
00	01	11	10	
①	–	2		0
	–	3	②	0
	–		③	1

Figure 9-15

If the circuit is in stable state ③, and the inputs change from $x_1 x_2 = 11$ to $x_1 x_2 = 10$, the circuit can return to stable state ② (the output changing from $Z = 1$ to $Z = 0$). If the circuit is in stable state ③, and the inputs change from $x_1 x_2 = 11$ to $x_1 x_2 = 01$, the circuit must change to a new stable state ④, for which $Z = 1$.

If the circuit is in stable state ④, and the inputs change from $x_1 x_2 = 01$ to $x_1 x_2 = 11$, the circuit can return to stable state ③. If the circuit is in stable state ④, and the inputs change from $x_1 x_2 = 01$ to $x_1 x_2 = 00$, the circuit can return to stable state ① (the output changing from $Z = 1$ to $Z = 0$). The primitive flow table at this stage of development is shown in Fig. 9-16.

$x_1 x_2$				Z
00	01	11	10	
①	–	2		0
1	–	3	②	0
	4	③	2	1
1	④	3	–	1

Figure 9-16

If the circuit is in stable state ①, and the inputs change from $x_1 x_2 = 00$ to $x_1 x_2 = 01$, the circuit must change to a new stable state ⑤, for which $Z = 0$.

If the circuit is in stable state ⑤, and the inputs change from $x_1 x_2 = 01$ to $x_1 x_2 = 00$, the circuit can return to stable state ①. If the circuit is in stable state ⑤, and the inputs change from $x_1 x_2 = 01$ to $x_1 x_2 = 11$, the circuit must change to a new stable state ⑥, for which $Z = 0$.

$x_1 x_2$

00	01	11	10	Z
①	5	–	2	0
1	–	3	②	0
–	4	③	2	1
1	④	3	–	1
1	⑤	6	–	0
–	5	⑥	2	0

Figure 9-17

If the circuit is in stable state ⑥, and the inputs change from $x_1 x_2 = 11$ to $x_1 x_2 = 01$, the circuit can return to stable state ⑤. If the circuit is in stable state ⑥, and the inputs change from $x_1 x_2 = 11$ to $x_1 x_2 = 10$, the circuit can return to stable state ②. The completed primitive flow table is shown in Fig. 9-17.

Construction of the state table or flow table forces the logic designer to completely account for all possible circuit action. There may have been certain input sequences that the designer had not initially considered. However, in the construction of the table, these sequences are called to his attention, and he must decide what the circuit action will be when these sequences occur (or else determine that the circuit action is optional).

Optional next state or output entries may arise in a table because certain transitions can never occur, or because we don't care what the circuit action is for a particular transition or state. Optional entries generally lead to greater circuit economy; therefore, the logic designer should not arbitrarily assign specific values to entries that could be optional.

PROBLEMS

1. Draw a state diagram and state table for the following circuit requirement: A sequential circuit is to have three pulse inputs x_1, x_2, and x_3 and two pulse outputs Z_1 and Z_2. The Z_1 pulse is to be coincident with the first x_2 pulse immediately following an x_1 pulse. The Z_2 pulse is to be coincident with all consecutive x_2 pulses immediately following an x_3 pulse.

2. Draw a state diagram and state table for the following circuit requirement: A sequential circuit is to have two pulse inputs x_1 and x_2 and one pulse output Z. The Z pulse is to be coincident with the second of two consecutive x_2 pulses immediately following exactly two consecutive x_1 pulses.

3. Draw a state diagram and state table for the following circuit requirement: A sequential circuit is to have two pulse inputs x_1 and x_2 and one pulse output Z. The Z pulse is to be coincident with the third and any further consecutive x_2 pulses immediately following exactly three consecutive x_1 pulses.

*4. Draw a state diagram and state table for the following circuit requirement: A sequential circuit is to have two pulse inputs x_1 and x_2 and one pulse output Z. The Z pulse is to be coincident with the third consecutive x_2 pulse immediately following three or more consecutive x_1 pulses.

5. Draw a state diagram and state table for the following circuit requirement: A sequential circuit is to have two level inputs x_1 and x_2 and one clock. A level output, Z, is to turn on with a clock pulse occurring with $x_1 x_2 = 01$.

Z is to turn off with the second of a sequence of clock pulses occurring with $x_1x_2 = 10$ immediately following a clock pulse occuring with $x_1x_2 = 01$. No other input sequence is to cause any change in output.

6. A sequential circuit is to have two level inputs x_1 and x_2 and one level output Z. The inputs represent, in binary, the numbers 0 through 3.

x_1x_2	*Number Representation*
00	0
01	1
10	2
11	3

If a change in input increases the represented number by *one*, the output is to turn on, if not already on. If a change in input decreases the represented number by *one*, the output is to change state. No other input sequence is to cause any change in output. All input changes are possible. Draw a primitive flow table for this circuit requirement.

7. A sequential circuit is to have two level inputs x_1 and x_2 and one level output Z. The inputs represent, in binary, the numbers 0 through 3. If a change in input increases the represented number, the output is to turn on, if not already on. If a change in input decreases the represented number, the output is to turn off, if not already off. No other input sequence is to cause any change in output. All input changes are possible except that both inputs will never turn off simultaneously. Draw a primitive flow table for this circuit requirement.

*8. A sequential circuit is to have two level inputs x_1 and x_2 and one level output Z. If the number of inputs that are *on* increases, the output is to turn off, if not already off. If the numbers of inputs that are *on* decreases, the output is to turn on, if not already on. No other input sequence is to cause any change in output. Both inputs will never turn off simultaneously; otherwise, all input changes are possible. Draw a primitive flow table for this circuit requirement.

10

State Table and
Flow Table Reduction

Having obtained the state table or flow table from the word statement of the problem, the next step in the synthesis procedure is to minimize the number of rows in the table. Each row must be defined by some combination of memory element states; minimizing the number of rows minimizes the number of memory element states that will be required. Of major importance, from a circuit economy standpoint, is that minimizing the number of memory element states minimizes the number of memory elements required. A secondary consequence is that, in general, the combinational logic is also reduced.

It should be stated that there are exceptions in which additional memory element states result in reduced combinational logic; even rarer are exceptions in which additional memory elements result in a great enough reduction in combinational logic to offset the cost of the added element. Even though there are occasional exceptions, minimizing the number of rows in the table most generally leads to the most economical circuit.

The aim is to obtain a minimum-row table that *covers* the original table in the sense that for all possible input sequences, the specified output sequence for both tables is the same.

There are three basic ways in which the number of rows in the table can be reduced:

1. Redundant states may have been introduced during construction of the table, it not being apparent that two or more states were equivalent. The table can be tested for equivalent states, and any redundant states, and therefore rows, eliminated.
2. If there are any optional entries in the table, next state or output state or both, it is possible that two or more states are "compatible," a term to be explained later. The table can be tested for compatible states, and the number of states, and therefore rows, may be able to be reduced.
3. In a state table, only one state may be assigned to each row. In a flow table, more than one stable state may be assigned to the same row, the input columns sharing in the specification. By a process called "merging," the number of rows in the flow table may be reduced by the assignment of more than one stable state to a row.

These means of table reduction will now be discussed. A table, obtained from the word statement of the problem, in which there are no optional entries is called a *completely specified table*; a table with any optional entries, next state or output state, is called an *incompletely specified table*. In a completely specified table, we discuss *equivalence*; in an incompletely specified table, we discuss *compatibility*.

Equivalent States

Assume a completely specified table. Two (stable) states are *equivalent* if, starting from either, any presented input sequence of arbitrary length results in an identical output sequence for both. The two states are also said to be *indistinguishable* since there is no way to distinguish which state is started from by observing the output sequence in response to any input sequence.

Conditions for equivalence are slightly different for state tables versus flow tables. In state tables, two states are equivalent if

1. The output states, pulse or level, associated with both states are the same,

and 2. For each possible input pulse there is a transition from these states to the same or equivalent states.

In flow tables, two stable states are equivalent if

1. They have the same input state (they are in the same column),

and 2. They have the same output state,

and 3. For each possible input change there is a transition from these stable states to the same or equivalent states.

Equivalent states can be replaced by a single state; equivalent rows are thus replaced by a single row. The single resulting row is said to *cover* the replaced equivalent rows, and the reduced table is said to cover the original table.

Following are some basic examples of equivalence (Figs. 10-1, 10-2, and 10-3). In all examples, only a portion of the table is shown, and in all examples, states 1 and 2 are equivalent and are replaced by the single state A.

	x_1	x_2
1	3/0	2/1
2	3/0	2/1

≡

	x_1	x_2
A	3/0	A/1

Figure 10-1

	x_1	x_2	Z_1Z_2
1	3	2	10
2	3	2	10

≡

	x_1	x_2	Z_1Z_2
A	3	A	10

Figure 10-2

x_1x_2: 00	01	11	10	Z_1Z_2
(1)	3	4	5	10
(2)	3	4	5	10

≡

x_1x_2: 00	01	11	10	Z_1Z_2
(A)	3	4	5	10

Figure 10-3

In the table in Fig. 10-4, we say that the pair of states 1 and 2 *implies* the pairs 4-5 (x_4 column) and 6-7 (x_5 column).

	x_1	x_2	x_3	x_4	x_5	Z
1	3	1	2	4	6	1
2	3	2	1	5	7	1

Figure 10-4

	x_1	x_2	x_3
1	3/0	1/1	2/0
2	3/0	2/1	1/0

≡

	x_1	x_2	x_3
A	3/0	A/1	A/0

Figure 10-5

A pair of states is equivalent only if all of the pairs it implies are equivalent; this is what constitutes the necessary condition of "transition to equivalent states." Thus, states 1 and 2 are equivalent only if states 4 and 5 are equivalent and states 6 and 7 are equivalent. It can also be said that the pair of states 1 and 2 implies itself (x_2 and x_3 columns), but equivalence is not conditional on self-implication; see, for example, Fig. 10-5.

Now for some slightly more involved examples. In Fig. 10-6, states 1 and 2 imply 3-4; states 3 and 4 are unconditionally equivalent; therefore, states 1 and 2 are equivalent.

	x_1	x_2	Z
1	5	4	0
2	5	3	0
3	4	6	1
4	3	6	1

≡

	x_1	x_2	Z
A	5	B	0
B	B	6	1

Figure 10-6

	x_1	x_2	Z
1	5	4	0
2	5	3	0
3	4	1	1
4	3	2	1

≡

	x_1	x_2	Z
A	5	B	0
B	B	A	1

Figure 10-7

In Fig. 10-7, states 1 and 2 imply 3-4; states 3 and 4 imply 1-2; there being no other conditions to satisfy, both equivalences are valid.

For larger tables, the determination of whether implied pairs are equivalent can involve a chain of interdependence, and the direct determination of equivalences can develop into a complex cyclic problem.

Implication Table

A systematic approach in determining equivalent states is to establish all *nonequivalences* first; all pairs of states not established as nonequivalent can then be made equivalent. An aid in this approach is the *implication table*. Its use will be demonstrated by an example (see the state table in Fig. 10-8).

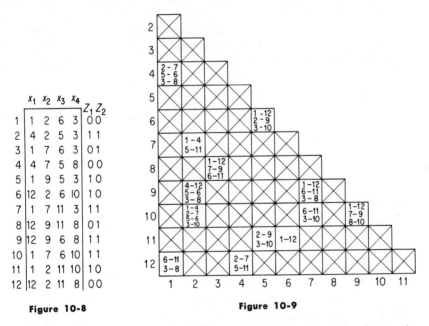

Figure 10-8 **Figure 10-9**

The implication table (Fig. 10-9) has a square for every pair of states in the state or flow table. If two (stable) states are nonequivalent because of output state conflicts, or because of input state conflicts in the case of flow tables, an X is entered in the corresponding square. If a pair of states has no such conflicts, all implied pairs (except itself) are entered in the corresponding square; if there are no implied pairs, the square is left blank.

A pair of states in the state or flow table for which each pair of entries is either identical or self-implying can be replaced by a single state before the start of construction of the implication table. If this is done, the implication table will, of course, be smaller and will have no blank squares.

Any further nonequivalences are found as follows. If an implied pair is nonequivalent, the implying pair is also nonequivalent. The X squares are therefore selected one at a time, and the table is searched for any implied pair corresponding to the X square. If any are found, an X is entered in the square containing the implied pair, signifying that that square also represents a nonequivalent pair. As each X square is considered, a second X, or any other desired mark, is entered in the square so that it will not be considered again. This process is continued until there are no single-X squares remaining in the table; at this time, the double-X squares represent all of the nonequivalences, and the squares with no X's represent all of the equivalences. Any implied pairs remaining at this time must correspond to other equivalences.

The significant steps in the completion of the implication table in Fig. 10-9 are as follows. The nonequivalent pair 3-10 generates the nonequivalences 2-10, 5-6, 5-11, and 7-10; 8-10 generates 9-10. 5-6 then generates 1-4 and 2-9; 5-11 generates 2-7 and 4-12. No other nonequivalences are generated, and the completed implication table is shown in Fig. 10-10. For simplicity, only one X per square is shown in the illustrations. The equivalent pairs and their single state replacements are

$$1\text{-}12 \equiv A$$
$$3\text{-}8 \equiv B$$
$$6\text{-}11 \equiv C$$
$$7\text{-}9 \equiv D$$

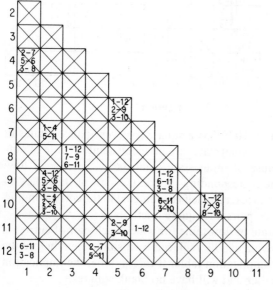

Figure 10-10

	x_1 x_2 x_3 x_4				$Z_1 Z_2$
A	A	2	C	B	0 0
2	4	2	5	B	1 1
B	A	D	C	B	0 1
4	4	D	5	B	0 0
5	A	D	5	B	1 0
C	A	2	C	10	1 0
D	A	D	C	B	1 1
10	A	D	C	10	1 1

Figure 10-11

The reduced table is shown in Fig. 10-11. Note that *all* occurrences of equivalent states are replaced by the new states.

Compatible States

Assume an incompletely specified table. Two (stable) states are *compatible* if, starting from either, any presented input sequence of arbitrary length results in an identical output sequence for both *wherever both are specified.*

Two (stable) states may be "equivalent" in all respects except for one or both of the following conditions:

1. For a given input, there is a transition from one of these states to a prescribed next state, whereas the transition from the second state is optional.
2. An output state associated with one of these states is prescribed, whereas for the second state, the corresponding output state is optional.

If either of the above conditions exists, the two states are said to be compatible.[1] Compatible states can be replaced by a single state (provided any implied compatibles are also replaced, in each case, by a single state). Where prescribed entries and optional entries correspond, the prescribed entry is retained in the covering row.

Some basic examples of compatibility follow. Again, in all examples, only a portion of the table is shown, and in all examples, states 1 and 2 are compatible and are replaced by the single state *A*.

(a) *Optional transition* (Fig. 10-12)

Note that if the optional entry were replaced with a 3, states 1 and 2 would be equivalent.

(b) *Optional transition* (Fig. 10-13)

	x_1	x_2
1	3/0	2/1
2	-/0	2/1

≡

	x_1	x_2
A	3/0	*A*/1

Figure 10-12

x_1x_2: 00	01	11	10	Z_1Z_2
(1)	3	-	5	10
(2)	-	4	5	10

≡

x_1x_2: 00	01	11	10	Z_1Z_2
(*A*)	3	4	5	10

Figure 10-13

(c) *Optional output* (Fig. 10-14)

(d) *Optional output* (Fig. 10-15)

	x_1	x_2
1	3/0	2/1
2	3/-	2/1

≡

	x_1	x_2
A	3/0	*A*/1

Figure 10-14

	x_1	x_2	Z_1Z_2
1	3	2	10
2	3	2	-0

≡

	x_1	x_2	Z_1Z_2
A	3	*A*	10

Figure 10-15

[1]The term *pseudo-equivalent* is also used.

(e) *Optional output* (Fig. 10-16)

x_1x_2: 00	01	11	10	Z_1Z_2
(1)	3	4	5	10
(2)	3	4	5	-0

≡

x_1x_2: 00	01	11	10	Z_1Z_2
(A)	3	4	5	10

Figure 10-16

There is an important difference between equivalent states and compatible states. If a state is equivalent to n other states, the n other states are necessarily all equivalent to each other, and the $n + 1$ states can be replaced by a single state. However, if a state is compatible with two other states, the two other states are not necessarily compatible with each other. An example is shown in Fig. 10-17. State 3 is compatible with states 1 and 2, but states 1 and 2 are incompatible with each other (they imply the pair 4-5, which is incompatible because of a conflict in output state). The compatible state pair 1-3 is replaced by state A, and the compatible state pair 2-3 is replaced by state B.

x_1x_2: 00	01	11	10	Z
(1)	4	–	7	0
(2)	5	6	–	0
(3)	–	6	7	0
1	(4)	6	–	1
2	(5)	6	–	0

x_1x_2: 00	01	11	10	Z
(A)	4	6	7	0
(B)	5	6	7	0
A	(4)	6	-	1
B	(5)	6	-	0

Figure 10-17

An important point to note is that an optional entry, next state or output state, need not always correspond to the same prescribed state. For example, in Fig. 10-17, in the combining of states 1 and 3, the optional entry in the $x_1x_2 = 01$ column of the 3 row corresponds to 4; in the combining of states 2 and 3, the same optional entry corresponds to 5. Theoretically, one could consider a row with an optional state as being replicated, one occurrence combining with each compatible row.

Only if n states are all pairwise compatible, each state compatible with each of the other states, can the n states be replaced by a single state.

The implication table is used to determine compatible states in an analogous manner to that for determining equivalent states: all incompatibles (X squares) are established; the squares with no X's then represent all of the pairwise compatibles. Any implied pairs correspond to other compatibles.

An example is shown in Fig. 10-18. It is suggested that the reader derive the implication table for practice.

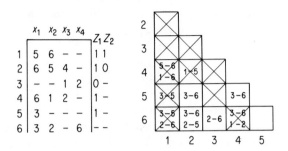

	x_1	x_2	x_3	x_4	$Z_1 Z_2$
1	5	6	–	–	1 1
2	6	5	4	–	1 0
3	–	–	1	2	0 –
4	6	1	2	–	1 –
5	3	–	–	–	1 –
6	3	2	–	6	– –

Figure 10-18

Compatible states, if they imply other compatible states, can be replaced by a single state only if the implied compatible states, in each case, are also replaced by a single state. The implied compatibles may, in turn, imply others, and so forth, a chain of implications being created, all of which must be satisfied if the initially considered compatibles are to be replaced by a single state. Remember that with the set of equivalent states in a completely specified flow table, there was no such dependency, all implied pairs being equivalent. Equivalence can be considered as a special case of compatibility.

Having obtained all of the pairwise compatibles, the objective now is to find the smallest closed and covering set of compatibles, each compatible to be replaced by a single state. A closed set of compatibles is one which includes all compatibles implied by the set; a covering set is one in which every state is included in at least one compatible. The minimal closed and covering set corresponds to a minimum-row table that covers the original table.

Unfortunately, there is no nonenumerative algorithm for finding the desired set. The set of maximal compatibles is helpful toward this end, however.

Maximal Compatibles

A *maximal compatible* is a compatible that is not included in any larger compatible. Two methods will be given for finding the maximal compatibles from the implication table; they will both be illustrated using the implication table in Fig. 10-19. Implied pairs are not pertinent to the methods; for that reason, the squares with no X's have all been left blank in the figure.

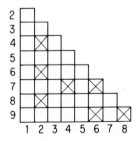

Figure 10-19

Method I

Start with the right-most column in the implication table, and move left until a column containing a pairwise compatible (X-free entry) is found. List all pairwise compatibles in that column.

EXAMPLE

Column 7:   78, 79

Move left to the next column containing a pairwise compatible. If the column designation is compatible with all members of any previously listed compatibles, then add the column designation to these compatibles. If the column designation is compatible with only a subset of some previously listed compatible, then add the column designation to that subset. Always choose a largest possible subset, that is, one not contained in a larger subset. Also add to the list any pairwise compatibles in that column that are not included in the compatibles already on the list. Eliminate, and do not add, any compatibles included in other compatibles.

EXAMPLE

Column 6:   78, 79, 68

Continue moving left, column by column, repeating the process, until the left-most column has been considered. The final list, along with any single states not included in any compatibles, comprises the maximal compatibles.

EXAMPLE

Column 5:   578, 579, 568
Column 4:   578, 579, 459, 4568˙
Column 3:   3578, 3579, 3459, 34568
Column 2:   3578, 23579, 3459, 34568
Column 1:   13578, 123579, 13459, 134568

Method II

Write every incompatible pair as a Boolean sum, and form the product of all of these sums.

EXAMPLE

$$(2 + 4)(2 + 6)(2 + 8)(4 + 7)(6 + 7)(6 + 9)(8 + 9)$$

Multiply out this expression to obtain an equivalent sum of products, eliminating all redundancy in the process.

EXAMPLE

$$2469 + 468 + 2678 + 279$$

For each resultant product, write the set of all missing states; these sets constitute all of the maximal compatibles.

EXAMPLE

13578, 123579, 13459, 134568

The maximal compatibles and all of their subsets comprise all possible compatibles. A minimal closed and covering set of these compatibles is desired. The implication table contains the information pertaining to implied pairs.

The set of maximal compatibles constitutes a closed and covering set, and represents an upper bound for the desired minimal set. (In a completely specified table, the maximal compatibles are mutually disjoint, and therefore all of them constitute a unique minimal set.)

The number of states in the largest *maximal incompatible* represents the lower bound for the minimal closed and covering set of compatibles. The maximal incompatibles can be found by interchanging the X and no-X squares in the implication table, and then using the methods for obtaining the maximal incompatibles.

While the literature presents various "methods" for obtaining the minimal closed and covering set of compatibles, the problem is extremely complex and at least partial enumeration is required.

Steps toward a solution often work at cross purposes. The larger a compatible, the more states it covers and, in general, the more implied pair requirements it satisfies. However, the larger a compatible, the more implied pairs it may specify, and therefore the more other compatibles that may be required.

A fairly simple example will be given. The state table and associated implication table are shown in Fig. 10-20. The maximal compatibles are obtained:

$$(1 + 2)(1 + 5)(2 + 3)(2 + 4)(4 + 5)(5 + 6) = 1246 + 1345 + 1346 + 25$$

Maximal compatibles: 35   26   25   1346

The largest maximal compatible, 1346, implies the three other maximal compatibles, resulting in an upper bound 4-state solution.

Noting that the states 1 and 4 appear only in the 1346 compatible, we can try various ways of splitting this compatible: 134; 146; 14; 13 and 46; or 16 and 34.

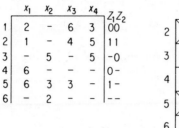
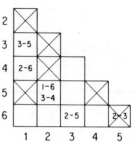

	x_1	x_2	x_3	x_4	Z_1Z_2
1	2	–	6	3	00
2	1	–	4	5	11
3	–	5	–	5	–0
4	6	–	–	–	0–
5	6	3	3	–	1–
6	–	2	–	–	––

Figure 10-20

1. 134 implies 26 and 35. A covering solution.
2. 146 implies 26; 35 must be added for a cover.
3. 14 implies 26; 35 must be added for a cover.

Of the preceding three 3-state solutions, we select solution 3 for its absence of redundancy.

4. 13, 46; 13 implies 35; 2 must be added for a cover. A 4-state solution.
5. 16, 34; nothing implied; 25 must be added for a cover. This is another good 3-state solution.

Using solutions 3 and 5, we obtain the two reduced state tables in Fig. 10-21.

$14 \equiv A$ $16 \equiv A$
$26 \equiv B$ $34 \equiv B$
$35 \equiv C$ $25 \equiv C$

	x_1	x_2	x_3	x_4	Z_1Z_2
A	B	–	B	C	00
B	A	B	A	C	11
C	B	C	C	C	10

	x_1	x_2	x_3	x_4	Z_1Z_2
A	C	C	A	B	00
B	A	C	–	C	00
C	A	B	B	C	11

Figure 10-21

In some of the literature, the replacing of equivalent or compatible states by a single state is called "merging," but we have reserved that term for its more common usage in the process to be described in the following section.

Merged Flow Table

State reduction ultimately results in a minimum-row *state* table, each state to be differentiated by some defining combination of memory element states. State reduction does not necessarily result in a minimum-row *flow* table, however. More than one stable state can occupy the same row, and a combination of memory element states is needed to define each row, but not each stable state. Stable states can be differentiated not only by memory element states but also by input states; stable states with the same memory element state can therefore be differentiated by input states.

Having obtained a primitive flow table with the minimum number of stable states, the next step therefore is to *merge* rows of the primitive table and obtain a *merged flow table*. Merging reduces the number of rows in the flow table by placing more than one stable state in a row.

In the primitive flow table, all transitions between stable states involve an input state change followed by a memory element state change. In the merged flow table, transitions between stable states in the same row are realized by input state changes only.

The rules for merging are as follows:

1. Two or more rows can be merged if they contain no conflicting state numbers in any column. For example, two rows can be merged if each column contains either two like state numbers, one state number and an optional entry, or two optional entries.
2. All state numbers appearing in the merging rows are written in the respective columns of the merged row. If a state number is circled in one of the merging rows, it is circled in the merged row, retaining the stable state designation.

The output states in no way affect merging. However, in the primitive flow table, the output state in each row corresponds to the stable state in that row. In the merged flow table, in which there is more than one stable state in a row, the relationship of each output state to its corresponding stable state must be maintained. To this end, the construction of a separate output state table corresponding, column and row, with the merged flow table could be started, and the output state corresponding to each stable state could be entered. Or the output state information could be moved adjacent to each stable state entry in the merged flow table.

The merged flow table may have to be modified during the memory element state assignment, however, and perhaps the most efficient way of handling the problem is simply to retain the primitive flow table, with its output state information, along with the merged flow table. Later on, after the memory element state assignment has been made, the output state information can be considered. This is the method that we shall follow.

As an example of a merger, the two rows in Fig. 10-22(a) are merged as shown in Fig. 10-22(b).

Generally, there is more than one way of merging the rows of a flow table, and the choice of mergers can affect circuit economy. In obtaining an optimum merger, a *merger diagram* is useful.

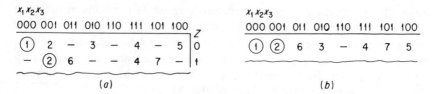

Figure 10-22

To construct a merger diagram, the stable state numbers are arranged in a basically circular array. The numbers are used here only to identify the rows of the primitive flow table. If, in the flow table, two rows can be merged, the corresponding stable state numbers in the merger diagram are connected by a line. All pairs of rows are examined for a possible merger, and after all connecting lines have been drawn, the merger diagram is inspected for the optimum way of merging. The aim, in general, is to merge so as to obtain the minimum number of rows in the merged flow table.

A primitive flow table and its associated merger diagram are shown in Fig. 10-23. Referring to the merger diagram, note that row 2 can merge with row 1 or row 3, but that rows 1 and 3 cannot merge with each other. Therefore, rows 1, 2, and 3 cannot all merge into one row, and a choice must be made between the merger of rows 1 and 2 and the merger of rows 2 and 3.

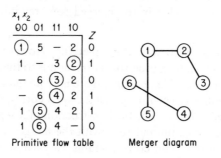

Primitive flow table   Merger diagram

Figure 10-23

A merger of rows 1, 2, and 5 cannot be made for the same reason, and a choice must be made between the merger of rows 1 and 2 and the merger of rows 1 and 5.

The mergers of rows 1 and 5, 2 and 3, and 4 and 6 result in a three-row flow table, which is optimum. A merger of rows 1 and 2 would not be desirable, since it would leave rows 3 and 5 unmerged, and the resulting flow table would contain four rows. The optimum three-row merged flow table is shown in Fig. 10-24.

Figure 10-25 illustrates three-row mergers. Rows 1, 2 and 6 can all merge into one row, as can rows 3, 4 and 5. Note that a four-row merger between

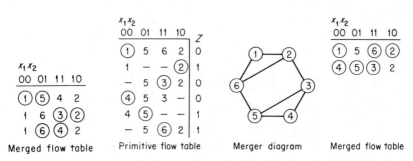

Figure 10-24   Figure 10-25

rows 2, 3, 5 and 6 is not possible, since rows 2 and 5 cannot merge and rows 3 and 6 cannot merge.

Figure 10-26 is an example of a four-row merger. Rows 1, 2, 5, and 6 can all merge into one row, and there is also a two-row merger between rows 3 and 4.

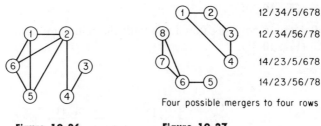

Figure 10-26   Figure 10-27

Often there may be more than one way of obtaining a minimum-row merger. In Fig. 10-27, there are four different ways of reducing to a four-row merged flow table. When there is more than one minimum-row merger, all of them should be considered, since there is no way of knowing at this stage of design which merger will result in the most economical circuit.

PROBLEMS

1. Draw a covering state table with the minimum number of states (Fig. 10-28).

2. Draw a covering flow table with the minimum number of stable states (Fig. 10-29).

	x_1	x_2	x_3	x_4	Z_1Z_2
1	5	2	11	1	11
2	3	2	11	12	01
3	3	7	6	12	00
4	5	7	4	1	00
5	5	2	4	1	10
6	10	2	6	12	00
7	3	7	11	1	01
8	10	8	4	12	01
9	5	8	6	9	11
10	10	7	6	12	10
11	10	8	11	9	00
12	10	7	11	12	11

Figure 10-28

x_1x_2

00	01	11	10	Z
①	6	9	11	0
②	6	7	10	0
③	4	8	10	0
3	④	9	11	1
1	⑤	9	12	1
1	⑥	8	10	1
1	5	⑦	11	0
3	6	⑧	11	0
1	4	⑨	10	0
3	6	9	⑩	1
1	4	8	⑪	1
2	4	8	⑫	0

Figure 10-29

*3. Draw a covering flow table with the minimum number of stable states (Fig. 10-30).

4. Draw a covering flow table with the minimum number of stable states (Fig. 10-31).

x_1x_2

00	01	11	10	Z
①	5	9	11	0
②	5	7	11	0
③	4	8	–	–
–	④	9	11	1
1	⑤	8	11	–
3	⑥	7	10	1
1	6	⑦	11	0
–	5	⑧	11	–
1	4	⑨	11	0
1	4	8	⑩	0
2	5	9	⑪	1

Figure 10-30

x_1x_2

00	01	11	10	Z_1Z_2
①	–	3	5	00
②	–	4	7	00
1	–	③	7	11
2	–	④	6	11
1	8	–	⑤	10
–	9	4	⑥	10
2	–	3	⑦	10
1	⑧	3	–	01
2	⑨	4	–	10

Figure 10-31

5. Obtain the maximal compatibles (Fig. 10-32).

6. Merge the primitive flow table in Fig. 10-33.

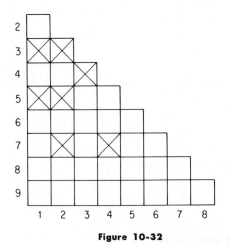

Figure 10-32

$x_1 x_2$

00	01	11	10	Z
①	4	2	–	0
–	–	②	5	0
8	③	–	5	0
1	④	–	7	1
8	4	2	⑤	0
8	3	⑥	–	1
1	4	–	⑦	1
⑧	3	6	–	1

Figure 10-33

*7. Merge the primitive flow table in Fig. 10-34.

$x_1 x_2$

00	01	11	10	Z
①	5	–	4	0
②	–	–	9	1
–	7	③	–	0
6	7	3	④	1
1	⑤	8	9	1
⑥	7	–	9	0
6	⑦	8	–	0
–	7	⑧	9	1
2	5	3	⑨	0

Figure 10-34

11

State Assignment

The minimum-row state table or flow table having been obtained, the next step is to assign a combination of memory element states, y's, to each state in the state table, or to each row in the flow table.

If there are r states in a minimum-row *state* table, n memory elements will be required, where

$$2^n \geq r > 2^{n-1}$$

With n memory elements, 2^n combinations of memory element states are available; this number may be equal to or greater than the number of states in the state table. A few examples of these relationships are expressed in the following table:

n	2^n	r
1	2	2
2	4	3–4
3	8	5–8
4	16	9–16

If there are r rows in a minimum-row *flow* table, n represents the *minimum* number of memory elements that will be required; more elements than this minimum may be necessary, due to considerations that will be discussed later in the chapter.

State assignment for flow tables is much more involved than that for state tables, and most of this chapter is devoted to flow table assignment. State assignment for state tables, being simpler, will be discussed first.

State Assignment for State Tables

For an r-state state table, which requires n memory elements, there are

$$\frac{2^n!}{r!(2^n - r)!}$$

ways of selecting r out of the 2^n possible combinations. For each of these ways, there are $r!$ permutations of assigning the r combinations to the r rows, making the total number of possible assignments

$$\frac{2^n!r!}{r!(2^n - r)!} = \frac{2^n!}{(2^n - r)!}$$

For each of these assignments there are 2^n ways of interchanging the *on* and *off* states of the memory elements and there are $n!$ ways of interchanging memory elements. There are thus

$$\frac{2^n!}{(2^n - r)!2^n \cdot n!} = \frac{(2^n - 1)!}{(2^n - r)!n!}$$

nontrivial assignments for an r-state state table.

Some values are tabulated below.

r	n	Number of Nontrivial Assignments
2	1	1
3	2	3
4	2	3
5	3	140
6	3	420
7	3	840
8	3	840
9	4	10,810,800

By way of example, the state assignment for a four-state table will be examined in detail. There are twenty-four possible assignments, but only three nontrivial variations. An arbitrary set of three nontrivial variations follows. These assignments will be utilized throughout the rest of the book, and are labeled $\#1$, $\#2$, and $\#3$ for reference.

	$\#1$ $y_1 y_2$	$\#2$ $y_1 y_2$	$\#3$ $y_1 y_2$
1	00	00	00
2	01	01	11
3	11	10	01
4	10	11	10

The "all memory elements off" state, in this case, $y_1 y_2 = 00$, is arbitrarily assigned to the "power on" state which we shall assume is state 1. For state 2, there is a choice of one or both memory elements on; $y_1 y_2 = 01$ is arbitrarily chosen for the "one element on" state. If state 2 is assigned $y_1 y_2 = 01$, there is a choice for state 3 of "the other element on," $y_1 y_2 = 10$, or both memory elements on. If state 2 is assigned $y_1 y_2 = 11$, state 3 must be assigned one element on; $y_1 y_2 = 01$ is arbitrarily chosen. State 4, in all cases, must be assigned the remaining memory element state combination.

The choice of state assignment is arbitrary; however, the choice can affect circuit economy. Unfortunately, there is no method of state assignment that guarantees minimum circuit cost. The assignment ultimately affects both the memory excitation logic and the circuit output logic. To further compound the problem, different types of memory elements differently influence the effect of the state assignment on the memory excitation logic. There is, therefore, no way to predict the worth of different assignments; they must be tried and evaluated. It helps, in this regard, for one to appreciate the difference between trivial and nontrivial variations in assignment.

Three examples of state assignment for state tables are shown in Figs. 11-1, 11-2, and 11-3. Note, in each table, that the same assignment is made for every occurrence of the same state number.

	x_1 x_2		$y_1 y_2$	x_1 x_2	
1	2/0	1/0	00	01/0	00/0
2	3/0	1/0	01	11/0	00/0
3	4/0	1/1	11	10/0	00/1
4	4/0	1/0	10	10/0	00/0

Assignment $\#1$

Figure 11-1

	x_1 x_2		$y_1 y_2$	x_1 x_2	
1	2/0	1/0	00	11/0	00/0
2	3/0	1/0	11	01/0	00/0
3	4/0	1/1	01	10/0	00/1
4	4/0	1/0	10	10/0	00/0

Assignment $\#3$

Figure 11-2

	x_1 x_2		Z
1	1	2	0
2	3	2	1
3	1	2	1

$y_1 y_2$	x_1 x_2		Z
00	00	01	0
01	11	01	1
11	00	01	1
10	--	--	--

Assignment $\#1$

Figure 11-3

One can think of state table assignment as follows: Each state number in the table is replaced by an assigned binary number of n digits; each position of the binary number relates to a memory element, and the corresponding digit, 1 or 0, describes the on or off state of the element.

With n memory elements, there will be 2^n combinations of memory element states in the state table with assignment. If the number of states, r, in the minimum-row state table is less than 2^n, there will be $2^n - r$ combinations of memory element states that never occur. In the state table with assignment, all entries corresponding to these spare combinations are optional. In Fig. 11-3, the memory element state $y_1 y_2 = 10$ is a spare state.

State Assignment for Flow Tables

We do not have the freedom to assign arbitrary combinations of memory element states to each row of a flow table in the same manner that assignments are made for state tables. In general, only a subset of the possible n-element assignments are applicable, and sometimes none are. In the latter case, assignments with more than n elements are required.

Let us trace some circuit action in a flow table with assignment. An assignment (not arbitrary) has been made for the flow table in Fig. 11-4. Starting with a stable state (say ①), a change in input state can cause a change to an unstable state (5 or 8) or to another stable state (②). A change in input state relates to horizontal movement in a row of the flow table. If the change were to an unstable state, a change to the corresponding stable state will follow (⑤ or ⑧). Vertical movement in a column of the flow table relates to a change in memory element state.

$y_1 y_2$	$x_1 x_2$ 00	01	11	10
00	①	②	5	8
01	3	2	⑤	⑥
11	③	④	7	6
10	1	4	⑦	⑧

Figure 11-4

Study of the flow table in this example will show that each transition from an unstable state to a stable state involves the change of only a single memory element state.

Races

In the assignment in Fig. 11-5, this is not the case. There are four transitions specified with this assignment that call for both memory elements to change simultaneously:

1 to ①	$y_1 y_2 = 11$ to 00
3 to ③	$y_1 y_2 = 01$ to 10
6 to ⑥	$y_1 y_2 = 10$ to 01
8 to ⑧	$y_1 y_2 = 00$ to 11

$y_1 y_2$	$x_1 x_2$ 00	01	11	10
00	①	②	5	8
01	3	2	⑤	⑥
11	1	4	⑦	⑧
10	③	④	7	6

Figure 11-5

Memory elements cannot be guaranteed to change exactly together; therefore, when multiple elements are called upon to change simultaneously, they may, in fact, change in any order. Examining the 1 to ① transition, for example, if both elements change exactly together, the transition is correctly y_1y_2 11 to 00. If element 2 responds before element 1, the transition is $y_1y_2 = 11$ to 10, and the circuit terminates incorrectly in stable state ③. If element 1 responds before element 2, the transition is $y_1y_2 = 11$ to 01, and the next circuit state is unstable state 3; a change to the corresponding stable state ③ will follow, and again the circuit terminates incorrectly.

When more than one memory element is called upon to change at the same time in level operation, a *race* is said to exist. If a race can terminate in any stable state other than the desired one, or can endlessly cycle (this will be explained later), the race is termed a *critical race*. All of the races in the preceding example were critical.

The behavior of a circuit with a critical race is not predictable, and critical races represent improper design and must be avoided. The flow table assignment problem is the problem of avoiding critical races.[1]

If a race, no matter what the relative memory element response times, always terminates in the desired stable state, the race is termed a *noncritical race*. An example of a noncritical race is shown in Fig. 11-6. Assume that the circuit is in stable state ④ and that the input state changes from $x_1x_2 = 01$ to 00. The circuit will enter unstable state 1, with $y_1y_2 = 11$. The specified transition, 1 to ①, then calls for the memory element change $y_1y_2 = 11$ to 00. If both elements change exactly together, the transition is correctly to stable state ①. If element 2 responds before element 1, the transition is $y_1y_2 = 11$ to 10; the circuit is then again in an unstable state 1. The specified transition, 1 to ①, now calls for the memory element change $y_1y_2 = 10$ to 00; this change involves only a single memory element, and the transition terminates correctly in stable state ①. If element 1 responds before element 2, the transitions will also be seen to terminate in stable state ①. Thus, no matter what the relative response times, the circuit always terminates correctly in stable state ①, and the race is noncritical.

x_1x_2

y_1y_2	00	01	11	10
00	①	②	5	8
01	1	2	⑤	⑥
11	1	④	7	6
10	1	4	⑦	⑧

Figure 11-6

Cycles

Before methods of flow table state assignment are discussed, one more type of transition will be examined. An unstable state does not always have

---

[1]Critical races may be avoided if the memory elements inherently, or by the insertion of proper delays, respond in a predictable order. The general case of indeterminate relative response times will be assumed, however.

to be directed to its corresponding stable state. If there are two or more similar unstable states in a column, only one must be directed to the corresponding stable state; the rest may be directed either to the stable state or to another similar unstable state. A transition directed from one unstable state to another is called a *cycle*, and is signified in the flow table by an arrow leading from the one to the other. The absence of an arrow leading from an unstable state indicates that that state is directed to its corresponding stable state. (The fact that a state is directed to another does not necessarily mean that the next state is the one to which the transition is directed, since noncritical races could be involved.)

A cycle is illustrated in the $x_1x_2 = 00$ column of the flow table in Fig. 11-7. If the circuit is in stable state ⑧, for example, and the input state changes from $x_1x_2 = 10$ to 00, there will be a cycle $y_1y_2 = 10$ to 11 to 01 to 00. If the circuit is in stable state ④, and the input state changes from $x_1x_2 = 01$ to 00, there will be a cycle $y_1y_2 = 11$ to 01 to 00.

y_1y_2 \ x_1x_2	00	01	11	10
00	①	②	5	8
01	1	2	⑤	⑥
11	1	④	7	6
10	1	4	⑦	⑧

Figure 11-7

Whereas critical races must be avoided, noncritical races and cycles are not only permissible but may be necessary for an optimum assignment. Noncritical races have the shorter transition times; cycles, on the other hand, are sometimes used to introduce additional delay.

A more complex example is shown in Fig. 11-8. If the circuit is in stable state ②, and there is an input change from $x_1x_2 = 01$ to 00, there will be a race from $y_1y_2y_3 = 111$ to 010. Depending on the outcome of the race, the next state may be $y_1y_2y_3 = 010$, 110, or 011. The 010 state changes to the stable 000 state. The 110 state cycles through 100 to the stable 000 state.

The 011 state attempts to change to 000: another race, the next state being 000 (stable), 010 or 001. The 010 and 001 states both change to the stable 000 state. Therefore, no matter what the outcome of the races, the circuit action eventually terminates in stable state $y_1y_2y_3 = 000$, and the

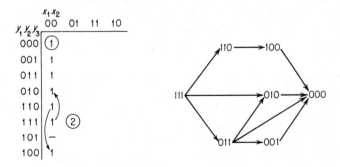

Figure 11-8

races are thus noncritical. All possible circuit actions are illustrated diagrammatically in Fig. 11-8.

The reader can benefit by studying in Fig. 11-9 all of the allowable assignments for the column shown, and convincing himself that no other assignment is valid. One invalid assignment is shown in Fig. 11-10, with all possible circuit actions illustrated diagrammatically. Note the critical race: if memory element 2 responds before memory element 1, the circuit will endlessly cycle between the $y_1 y_2 = 10$ and 11 states.

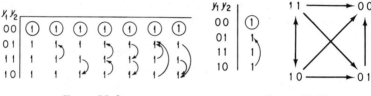

Figure 11-9   Figure 11-10

Transition Map

In making a state assignment for flow tables, an attempt is, of course first made to use no more than the minimum number of memory elements. A *transition map* is helpful in deriving assignments (see Fig. 11-11).

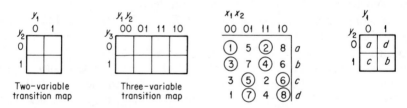

Figure 11-11   Figure 11-12

Each square in the transition map represents a combination of memory element states. As a first step in using a transition map, each row in the flow table is assigned a letter label. The assignment of a combination of memory element states to a row in the flow table is denoted by an entry of the row label in the corresponding square of the transition map. See the example in Fig. 11-12.

With n memory elements, if the number of rows, r, in the minimum-row flow table is less than 2^n, there will be $2^n - r$ spare rows. In any particular column, the entries in these spare rows can be utilized to avoid critical races; they can be used, for example, in cycles or noncritical races. Any entries not used are optional. In general, the more optional entries, the more economical the resulting circuit; therefore, spare entries should be used with discretion.

If there is but a single occurrence of an unstable state, and there are no spare entries in that column, there must be a direct transition from the unstable to the stable state; that is, there must be a change in only a single memory element state. The row with the unstable state and the row with the corresponding stable state must therefore be assigned combinations of memory element states differing in only one variable. States differing in only one variable are represented on the transition map by logically adjacent entries.

Arbitrarily, the "power on" state, which we shall assume is stable state ①, will be placed in the first row, first column, of the flow table, and will be assigned the all-0 state: all input states equal 0 and all memory element states equal 0. The assignment of $y_1y_2 = 00$ to row a is denoted by placing an a entry in the $y_1y_2 = 00$ square of the transition map (see Fig. 11-12).

The flow table is first examined for any required direct transitions between row a and another row, that is, from row a to another row, or from another row to row a. If any are found, the row labels of these other rows must be placed on the transition map in squares adjacent to the a entry. In an n-variable transition map, there are n such adjacent squares. If more adjacencies are needed, it signifies that more than n memory elements will be required.

In the flow table in Fig. 11-12, a direct transition is required between rows

a and c (stable states ① or ② to ⑤)

a and d (stable states ① or ② to ⑧)

a and c (stable states ⑤ or ⑥ to ②)

a and d (stable states ⑦ or ⑧ to ①)

The c and d row labels must therefore be placed on the transition map in the squares adjacent to the a entry: $y_1y_2 = 01$ or 10 (the state assignment for rows c and d must differ in only one variable from the assignment for row a). Rows c and d are arbitrarily assigned as shown in Fig. 11-12; the reverse assignment is a trivial variation.

Row b can have only the remaining assignment $y_1y_2 = 11$. However, the entire table must be examined for all required direct transitions before it is known whether the total assignment can be valid. The remaining required direct transitions are between rows.

b and c (stable states ③ or ④ to ⑥)

b and d (stable states ③ or ④ to ⑦)

b and c (stable states ⑤ or ⑥ to ③)

b and d (stable states ⑦ or ⑧ to ④)

All direct transitions are therefore represented on the transition map by adjacent entries, and the assignment is valid. The flow table with assignment is shown in Fig. 11-13. The row labels are retained for instructional purposes. The reflected ordering is used for the row assignments, as with the columns, preparatory to obtaining memory excitation and circuit output maps from the flow table. Note that this may call for a rearrangement in the order of the rows, as in this example.

y_1y_2 \ x_1x_2	00	01	11	10	
00	①	5	②	8	*a*
01	3	⑤	2	⑥	*c*
11	③	7	④	6	*b*
10	1	⑦	4	⑧	*d*

Figure 11-13

x_1x_2: 00	01	11	10	
①	②	3	④	*a*
⑤	⑥	3	⑦	*b*
1	6	③	10	*c*
⑧	⑨	3	⑩	*d*

Figure 11-14

An attempt to make a two-element assignment for the flow table in Fig. 11-14 will not be successful. Direct transitions are required between rows

c and *a* (stable state ③ to ①)
c and *b* (stable state ③ to ⑥)
c and *d* (stable state ③ to ⑩)

requiring *a*, *b*, and *d* to be adjacent to *c*. This is an impossibility, and an assignment for this flow table, free of critical races, will require three memory elements.

The left three columns of the flow table in Fig. 11-15 lead to the transition map shown in the figure. The transition between rows *a* and *c* (stable state ① to ④) in the $x_1x_2 = 10$ column does not invalidate the transition map assignment since this transition need not be direct. There are two occurrences of unstable state 4 in the column, and a critical race can be avoided by the

x_1x_2: 00	01	11	10	
①	2	3	4	*a*
5	②	⑥	⑦	*b*
⑤	8	6	④	*c*
1	⑧	③	4	*d*

y_2 \ y_1	0	1
0	*a*	*d*
1	*b*	*c*

Figure 11-15

y_1y_2 \ x_1x_2	00	01	11	10
00	①	2	3	4
01	5	②	⑥	⑦
11	⑤	8	6	④
10	1	⑧	③	4

Figure 11-16

cycle a to d to c for this transition. The flow table with assignment is shown in Fig. 11-16.

A column having only one stable state need not be of any concern in making an assignment, since cycles and noncritical races can be prescribed in such a column (see, for example, Fig. 11-9).

The utilization of spare entries will be illustrated with an example. An assignment for a three-row flow table can always be achieved with two memory elements, although the spare fourth state must sometimes be utilized to avoid critical races. This requirement occurs when there are transitions between all three pairs of rows, a and b, a and c, and b and c, as illustrated in Fig. 11-17. No matter how assignments are made, a transition will be required between two rows whose assignments differ in two variables.

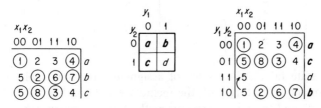

Figure 11-17   **Figure 11-18**

With the assignment shown in Fig. 11-18, the state assignments of rows b and c differ in two variables; however, transitions between these two rows can be prescribed to cycle through the spare state, labeled d, as shown in the flow table with assignment.

Two other assignments for the same problem are shown in Figs. 11-19 and 11-20, the first in which a and c are not adjacent, and the second in which a and b are not adjacent. The three assignments generally lead to different solutions.

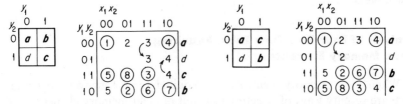

Figure 11-19   **Figure 11-20**

It can be shown that a flow table with $r = 2^n$ rows, called a full flow table, can always be realized, free of critical races, with no more than $2^n - 1$ memory elements. A flow table with $r = \frac{3}{4} \cdot 2^n$ rows, called a three-quarter flow table, can always be realized with no more than $2n - 2$ memory elements. Some values are shown in the following table:

Number of Rows in Flow Table, r	Lower Bound on Number of Memory Elements Required, n	Full Flow Table, $r = 2^n$	3/4 Flow Table, $r = \frac{3}{4} \cdot 2^n$	Upper Bound on Number of Memory Elements Required	
				$2n - 1$	$2n - 2$
2	1	✓		1	
3	2		✓		2
4	2	✓		3	
6	3		✓		4
8	3	✓		5	
12	4		✓		6
16	4	✓		7	
24	5		✓		8
32	5	✓		9	

As in state tables, the choice of assignment for flow tables can affect circuit economy, affecting both the memory excitation logic and the circuit output logic, and there is no general method of assignment that guarantees minimum circuit cost. As with state tables, there is no way to predict the worth of different assignments; they must be tried and evaluated.

While avoiding critical races is the designer's primary concern, he should, at the same time, exploit the allowable variations in assignment to the fullest extent and not unnecessarily restrict himself. On the following pages is presented an analysis of four-row flow table requirements necessitating three memory elements. Study of this analysis will give the reader a general appreciation for the trivial versus nontrivial variations in assignment, and will give him a feel for both the flexibility and restrictions involved in making flow table assignments. The concepts discussed can be extended to larger flow tables.

Analysis of Four-Row Flow Table Requiring Three Memory Elements

When three memory elements are required for a four-row flow table, there are seventy ways of selecting four out of eight memory element states for assignment to four rows:

$$_8C_4 = \frac{8!}{4!4!} = 70$$

Analysis shows that each of the seventy combinations falls into one of six different patterns. A word description of each pattern, with an example of each in a transition map, is given on p. 172. The selected states are arbitrarily labeled w, x, y, and z.

The patterns are numbered for reference purposes only. It should not be inferred that in a particular problem, some pattern should be "chosen"; the assignment should most generally be "tailor-made" for each problem. The application of pattern #1 is limited; it can be used for any[2] four-row flow table in which there are transitions between five or fewer of the six pairs of rows, but it can be used only for *some* cases in which there are transitions between all six pairs of rows. Patterns #2 through #6 can be used for any[2] four-row flow table.

For each combination, there are twenty-four permutations of row assignment:

$$_4P_4 = 4! = 24$$

that is, there are twenty-four ways of assigning four selected states to four rows. Analysis shows that for each pattern, certain permutations lead to solutions that are equivalent with memory elements interchanged. The number of nonequivalent row assignments for each pattern follows:

Pattern	Number of Nonequivalent Row Assignments
#1	3
#2	4
#3	12
#4	12
#5	6
#6	1
	38

If three memory elements are required for a four-row flow table, there are thus thirty-eight secondary assignments that will, in general, lead to different solutions (thirty-five in cases in which pattern #1 is not applicable).

Examples of the thirty-eight assignments are shown in Fig. 11-27. The pattern variations in the fourth example of pattern #2, and in the last six examples of patterns #3 and #4, are made to retain the $y_1y_2y_3 = 000$ assignment for row a.

Each pattern will now be examined in more detail and illustrated by a flow table example. For each pattern, a resulting flow table with assignment is shown for study.

---

[2]Exceptions can occur when cycles are required as part of the original circuit output specifications; see *Transient Outputs; Cyclic Specifications* in Chapter 13. Cyclic specifications are not considered in the discussion that follows.

Pattern	*Word Description*	*Example*
#1	w and x differ in one variable x and y differ in one variable y and z differ in one variable z and w differ in one variable	 **Figure 11-21**
#2	x and w differ in one variable x and y differ in one variable x and z differ in one variable	 **Figure 11-22**
#3	w and x differ in one variable x and y differ in one variable y and z differ in one variable z and w differ in three variables	 **Figure 11-23**
#4	x and w differ in one variable x and y differ in one variable x and z differ in three variables	 **Figure 11-24**
#5	w and x differ in one variable y and z differ in one variable w and y differ in three variables x and z differ in three variables	 **Figure 11-25**
#6	All pairs of states differ in two variables	 **Figure 11-26**

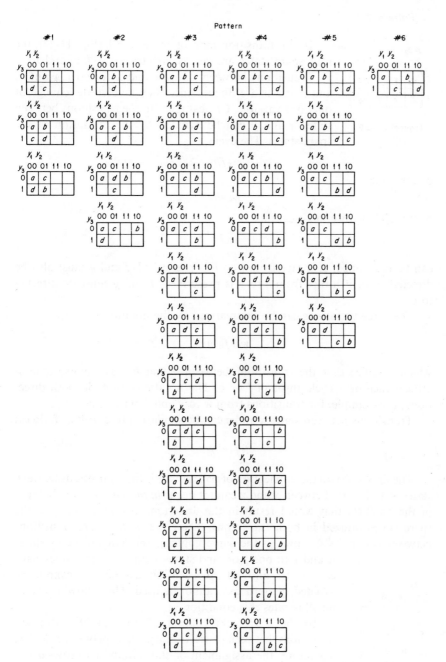

Examples of three−variable memory element for four−row flow tables

Figure 11-27

Pattern #1

$x_1 x_2$

y_3	00	01	11	10
0	*w*	*x*	*f*	*e*
1	*z*	*y*	*h*	*g*

Figure 11-28

In the transition map of pattern #1 (Fig. 11-28), the four spare states are labeled arbitrarily, $e, f, g,$ and h.

Transitions between states w and x, x and y, y and z, and z and w are direct, since they involve the change of only one variable. Critical races in the transitions between states w and y can be avoided by the utilization of the spare states. The cycle

$$wefhy$$

or the cycle

$$weghy$$

or the cycle with noncritical race

$$wehy$$

can be prescribed. If a race from e to h is prescribed, f and g must also be directed to h; if a race from h to e is prescribed, f and g must be directed to e.

The three variations can be summarized by the notation

$$we(fg)hy$$

which signifies that the transitions between e and h may involve a non-critical race or a cycle through f or g. The notations apply in both directions; for example, for transitions from w to y and from y to w.

Transitions between states x and z can be similarly prescribed as follows:

$$xf(eh)gz$$

The limitation on the application of pattern #1 is that transitions between states w and y and between x and z cannot occur *in the same column* because of the conflicts that would result in the direction, in that column, of the spare states utilized in both transitions. Therefore, if there are transitions between two pairs of rows in the same column, the assignment of states w and y to one pair, and states x and z to the other pair, cannot be allowed. If such an assignment cannot be avoided, pattern #1 cannot be used. The following example illustrates this condition.

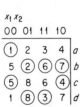

$x_1 x_2$

00 01 11 10

In the $x_1 x_2 = 00$ column of Fig. 11-29, there are transitions between rows a and d, and between b and c. Therefore, the assignment of states w and y to rows a and d, and the assignment of states x and z to rows b and c, or vice versa, is not allowed. In the $x_1 x_2 = 01$ column,

Figure 11-29

there are transitions between rows *a* and *b*, and between *c* and *d*. Therefore, the assignment of states *w* and *y* to rows *a* and *b*, and the assignment of states *x* and *z* to rows *c* and *d*, or vice versa, is not allowed. In the $x_1x_2 = 11$ column, there are transitions between the same pairs of rows as in the $x_1x_2 = 00$ column. In the $x_1x_2 = 10$ column, there are transitions between rows *a* and *c* and between *b* and *d*. Therefore, the assignment of states *w* and *y* to rows *a* and *c*, and the assignment of states *x* and *z* to rows *b* and *d*, or vice versa, is not allowed.

There is thus no assignment that will avoid transitions between states *w* and *y* and between *x* and *z* in the same column, and pattern #1 cannot be used for this flow table. In the example that follows, pattern #1 can be used.

The $x_1x_2 = 00$ and $x_1x_2 = 01$ columns in Fig. 11-30 are the same as in the preceding example, and therefore the assignment of states *w* and *y* to rows *a* and *d*, and the assignment of states *x* and *z* to rows *b* and *c*, or vice versa, is not allowed; and the assignment of states *w* and *y* to rows *a* and *b*, and the assignment of states *x* and *z* to rows *c* and *d*, or vice versa, is also not allowed.

x_1x_2: 00	01	11	10	
(1)	2	(3)	4	*a*
5	(2)	(6)	(7)	*b*
(5)	8	(9)	(4)	*c*
1	(8)	6	(10)	*d*

Figure 11-30

All six possible row pair transitions occur in the flow table, but the remaining two transitions occur in different columns: the transition between rows *b* and *d* occurs in the $x_1x_2 = 11$ column, and the transition between rows *a* and *c* occurs in the $x_1x_2 = 10$ column. Therefore, states *w* and *y* can be assigned to rows *b* and *d*, and states *x* and *z* can be assigned to rows *a* and *c*, or vice versa, and pattern #1 is thus applicable. The assignment in Fig. 11-31 is chosen. Transitions between rows *a* and *c* are prescribed by

$y_3 \backslash y_1y_2$	00	01	11	10
0	*a*	*b*	*f*	*e*
1	*d*	*c*	*h*	*g*

Figure 11-31

$$ae(fg)hc$$

and transitions between rows *b* and *d* are prescribed by

$$bf(eh)gd$$

A resulting flow table with assignment is shown in Fig. 11-32. The cycle *dgefb* is chosen for the *d* to *b* transition in the $x_1x_2 = 11$ column. The cycle with noncritical race *aehc* is chosen for the *a* to *c* transition in the $x_1x_2 = 10$ column. The choices of optional transitions in this and the following examples are arbitrary, and are selected to illustrate the various types of transitions possible.

The flow table in Fig. 11-33 will be used as a running example in the examination of patterns #2 through #6. The assignment illustrating each pattern is chosen arbitrarily.

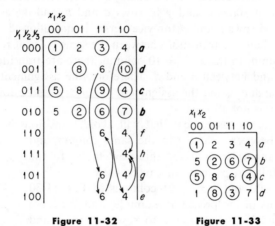

Figure 11-32   Figure 11-33

Pattern #2

See Fig. 11-34. Transitions between rows b and a, b and c, and b and d are direct. The remaining transitions are prescribed by

$$aec$$
$$afd$$
$$cgd$$

See Fig. 11-35.

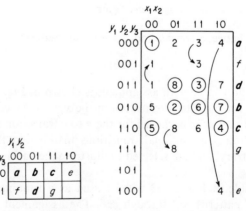

Figure 11-34   Figure 11-35

Pattern #3

See Fig. 11-36. Transitions between rows a and b, b and c, and c and d are direct. The remaining transitions are prescribed by

	00	01	11	10
0	a	b	c	e
1	f	g	d	h

$x_1 x_2$ (top), y_3 (left)

$$aec$$

$$bgd$$

$$a(ef)hd \quad \text{or} \quad af(gh)d$$

Figure 11-36

If the a to d transition $afhd$ is chosen, there can, of course, be no conflicts. However, the spare states e and g can be utilized in more than one type of transition without conflicts. For example, consider the transitions involving the spare state e: aec and $aehd$. Transitions from a to c and from a to d cannot occur in the same column. Transitions from c to a and from d to a, even if they do occur in the same column, cause no conflict, e being directed to a in both transitions. Also, of course, transitions *from* a and *to* a cannot occur in the same column.

When there are two types of transitions, with a row of a flow table common to both, for example, row a in the preceding example, the same spare states can be utilized in both transitions without conflict.

If two types of transitions have no rows of a flow table in common, the same spare states can be utilized in both transitions, without conflict, only if the transitions do not occur in the same column (see examples in the discussion of pattern #1).

When there is a choice of transitions between two rows, the one selected for any particular column is independent of that chosen in any other column. With pattern #3, for example $aehd$ might be chosen for the a to d transition in one column, and $afgd$ might be chosen in another column.

In the flow table in Fig. 11-37, the cycle with noncritical race dfa is chosen for the d to a transition in the $x_1 x_2 = 00$ column; and the cycle $aehd$ is chosen for the a to d transition in the $x_1 x_2 = 11$ column.

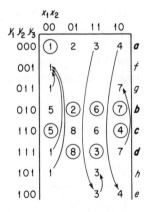

Figure 11-37

Pattern #4

See Fig. 11-38. Transitions between rows a and b, and b and c are direct. The remaining transitions are prescribed by

$$aec$$
$$a(ef)d$$
$$c(eh)d$$
$$bg(fh)d$$

Note that the spare states e, f and h can be utilized in more than one type of transition, similar to the spare states in pattern #3.

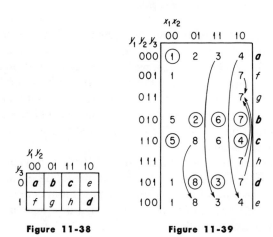

Figure 11-38   Figure 11-39

In the flow table in Fig. 11-39, the noncritical race da is chosen for the d to a transition in the $x_1 x_2 = 00$ column; the cycle aed is chosen for the a to d transition in the $x_1 x_2 = 11$ column; the cycle ced is chosen for the c to d transition in the $x_1 x_2 = 01$ column; and the cycle with noncritical race dgb is chosen for the d to b transition in the $x_1 x_2 = 10$ column.

Pattern #5

See Fig. 11-40. Transitions between rows a and b, and c and d are direct. The remaining transitions are prescribed by

Figure 11-40

$$a(ef)d$$
$$b(gh)c$$
$$aehc \quad \text{or} \quad afgc,$$
$$bhed \quad \text{or} \quad bgfd$$

All spare states can be utilized in more than one type of transition, subject to the following restriction: since transitions between rows a and c and between rows b and d involve no rows in common, if both types of transitions occur in the same column, the same spare states cannot be utilized in both, and the choice of transitions must be restricted to

$a(ef)d$		$a(ef)d$
$b(gh)c$	or	$b(gh)c$
$aehc$		$afgc$
$bgfd$		$bhed$

In the running example, the above restriction must be observed since, in the $x_1 x_2 = 10$ column, there is a transition from row a to c, and from d to b.

In the flow table in Fig. 11-41, the noncritical races da and bc are chosen for the respective d to a and b to c transitions in $x_1 x_2 = 00$ column; the cycles afd and chb are chosen for the respective a to d and c to b transitions in the $x_1 x_2 = 11$ column; and the cycles $aehc$ and $dfgb$ are chosen for the respective a to c and d to b transitions in the $x_1 x_2 = 10$ column.

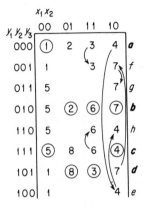

Figure 11-41

Pattern #6

See Fig. 11-42. The transitions are prescribed as follows:

$$a(eg)b$$
$$a(fg)c$$
$$a(ef)d$$
$$b(gh)c$$
$$b(eh)d$$
$$c(fh)d$$

y_3 \ $x_1 y_2$

	00	01	11	10
0	a	g	b	e
1	f	c	h	d

Figure 11-42

Note that all spare states can be utilized in more than one type of transition. Also note that noncritical races can be used for all transitions; this assignment might be selected for its short transition times.

In the flow table in Fig. 11-43, the cycles dea and bgc are chosen for the respective d to a and b to c transitions in the $x_1 x_2 = 00$ column. The cycle aeb and the noncritical race cd are chosen for the respective a to b and c to d transitions in the $x_1 x_2 = 01$ column. The noncritical races ad and cb are chosen for the respective a to d and c to b transitions in the $x_1 x_2 = 11$ column.

$x_1 x_2$

$x_1 y_2 y_3$	00	01	11	10	
000	①	2	3	4	a
001		8	3	4	f
011	⑤	8	6	④	c
010	5		6	4	g
110	5	②	⑥	⑦	b
111		8	6		h
101	1	⑧	③	7	d
100	1	2	3	7	e

Figure 11-43

The noncritical race *ac* and the cycle *deb* are chosen for the respective *a* to *c* and *d* to *b* transitions in the $x_1x_2 = 10$ column.

Assignment of Multiple Memory Element States to a Row

The discussion so far has been limited to assignments in which each row of a flow table is assigned *one* combination of memory element states. When the use of spare states is required, however, rows of a flow table can be assigned more than one combination of memory element states. We can think of a row as being replicated, each occurrence being assigned a different state. This type of assignment will now be examined.

An assignment for the three-row flow table previously examined (repeated in Fig. 11-44), requires the use of the spare fourth state. In a previous example, the assignment in Fig. 11-45 was made. In the transitions between rows *a* and *c*, critical races were avoided by cycles through the spare $y_1y_2 = 01$ state, labeled *d*, being prescribed.

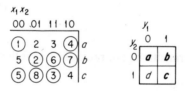

Figure 11-44   Figure 11-45

The preceding assignment can be modified by the assignment of the spare $y_1y_2 = 01$ state to row *a* or row *c*, which still avoids critical races in the transitions between rows *a* and *c*.

EXAMPLE

Figure 11-46

When more than one combination of memory element states is assigned to the same row, rows with equivalent stable states are created in the flow table. Subscripts are used to differentiate these rows and equivalent stable states for transition indentification. The output states associated with equivalent stable states are the same, and the selection of equivalent states for a particular transition is based fundamentally on the avoidance of critical races. Refer to Fig. 11-46.

Transitions from *b* to *a* are directed to row a_1; transitions from *c* to *a* are directed to a_2; transitions from a_2 to *b* cycle through a_1; and transitions from a_1 to *c* cycle through a_2.

The flow table with assignment is shown in Fig. 11-47.

Subscripts are used on unstable state numbers where necessary to specify to which of equivalent stable states a transition occurs.

Figures 11-48 and 11-49 show another assignment for the same problem, with the $y_1y_2 = 01$ secondary state assigned to row c instead of row a.

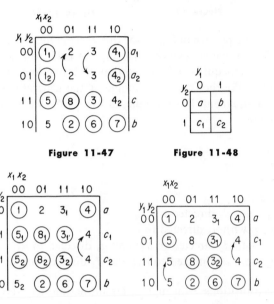

Figure 11-47

Figure 11-48

Figure 11-49

Figure 11-50

The assignment of multiple memory element states to a row can be modified by the replacement of some equivalent stable states with the corresponding unstable states. For example, the flow table in Fig. 11-49 could be modified as shown in Fig. 11-50. Note that all columns do not have to be treated in the same manner.

When three memory elements are required for a four-row flow table, and the assignment of more than one memory element state to a row is considered, many additional patterns become possible. Some examples follow (Figs. 11-51 to 11-55), in which all eight memory element states are assigned to rows.

y_3	y_1y_2			
	00	01	11	10
0	a_1	b_1	c_1	c_2
1	a_2	b_2	d_1	d_2

Figure 11-51

y_3	y_1y_2			
	00	01	11	10
0	a_1	b	c_1	c_2
1	a_2	d_1	d_2	d_3

Figure 11-52

y_3	y_1y_2			
	00	01	11	10
0	a	b	c_1	c_2
1	d_1	d_2	d_3	d_4

Figure 11-53

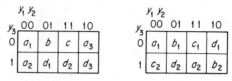

Figure 11-54   **Figure 11-55**

Notice that the pattern in Fig. 11-55 differs from the others in that no rows with equivalent stable states are adjacent, but each row is adjacent to one of each type of nonequivalent row, so that all transitions are direct.

One should take the fullest advantage of the generally wide selection of variations in assignment. For example, in Fig. 11-53, the possible transitions from d_1 to c are

1. $d_1 d_4 c_2$.
2. $d_1 d_4 d_3 c_1$.
3. $d_1 d_2 d_3 c_1$.
4. $d_1 d_2 d_3 d_4 c_2$.
5. $d_1 d_4 c_1$, with d_3 directed to c_1.
6. $d_1 d_2 d_3 c_2$, with d_4 directed to c_2.
7. $d_1 d_3 c_1$, with d_2 directed to d_3; and d_4 directed to d_3, c_1 or c_2.
8. $d_1 d_3 c_2$, with d_2 directed to d_3; and d_4 directed to c_2.

Note that in transitions 5 through 8, circuit action may terminate in either c_1 or c_2.

Flow Table to State Table

Once an assignment has been made for a flow table, the table becomes a state table, since each next state is now prescribed. To complete the assignment, each next state must be assigned the proper combination of memory element states. An example is shown, using the flow table with assignment in Fig. 11-56.

$y_1 y_2$ \ $x_1 x_2$	00	01	11	10
00	①	2	3_1	④
01	⑤	7	③₁	4
11	5	⑦	③₂	4
10	5	②	⑥	4

Figure 11-56

$y_1 y_2$ \ $x_1 x_2$	00	01	11	10
00	00			00
01	01		01	
11		11	11	
10		10	10	

Figure 11-57

For each stable state, the next memory element state is the same as the present memory element state. It is convenient to make the assignments for the stable states first (Fig. 11-57).

For each unstable state, the next memory element state combination is specified. Take, for example, the unstable state 5 in the $x_1x_2 = 00$ column, $y_1y_2 = 10$ row. The arrow indicates that the next state is the unstable state 5 in the $y_1y_2 = 11$ row; the next memory element state combination is therefore $y_1y_2 = 11$, which is assigned to the unstable state 5 in the $y_1y_2 = 10$ row.

For the unstable state 5 in the $y_1y_2 = 11$ row of this same column, the next state is the stable state ⑤; the next memory element state combination is therefore $y_1y_2 = 01$, which is assigned to this unstable state 5.

The assignments for the rest of the unstable states are made in a similar manner. The completed state table with assignment is shown in Fig. 11-58.

The *flow* table with assignment (e.g., Fig. 11-56) is retained; it will be used later, along with the primitive flow table, to obtain the output map and expressions.

y_1y_2 \ x_1x_2	00	01	11	10
00	00	10	01	00
01	01	11	01	00
11	01	11	11	00
10	11	10	10	00

Figure 11-58

PROBLEMS

1. Make an assignment for the flow table in Fig. 11-59.

*2. Make an assignment for the flow table in Fig. 11-60.

x_1x_2

00	01	11	10
①	2	5	⑧
③	4	⑤	7
3	②	⑥	7
1	④	5	⑦

Figure 11-59

x_1x_2

00	01	11	10
①	②	4	7
6	③	④	8
⑥	3	⑤	⑦
6	2	4	⑧

Figure 11-60

3. Make an assignment for the flow table in Fig. 11-14.

4. List all of the possible transitions from a_3 to d in Fig. 11-54. Indicate any directed transitions necessitated by races. Indicate where circuit action may terminate other than where prescribed.

5. Complete the assignment, assigning each next state the proper combination of memory element states (Fig. 11-61).

*6. Complete the assignment, assigning each next state the proper combination of memory element states (Fig. 11-62).

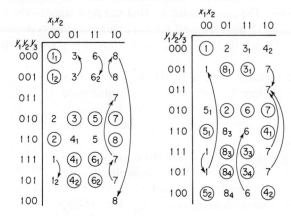

Figure 11-61   **Figure 11-62**

12

Excitation Maps
and Expressions

Let us briefly recapitulate what has been accomplished so far in the synthesis procedure for a sequential switching circuit. From the word statement of the problem, a state table, for pulse operation, or a flow table, for level operation, was constructed. The number of rows in the table was minimized, and a combination of memory element states was assigned to each state in the table.

Once this assignment is made, the required number of memory elements is known. Furthermore, given any present combination of memory element (y) states and input (x) states, the next combination of memory element states is specified. To realize these next memory element states, the present states of the y's and x's must properly excite the memory element inputs. The type of excitation varies with the type of memory element employed.

Figure 12-1 shows a schematic diagram of a memory element with its inputs and outputs. For each allowable memory input (W) excitation, with either present state of the memory element, $y = 0$ or 1, the next memory element state is a characteristic of the type of memory element. Given, then, a present state and desired next state of the element, the characteristics of the element prescribe the required memory excitation.

Figure 12-2 shows a portion of the schematic diagram of a generalized sequential circuit (from Fig. 8-1). Given a present combination of x and y states, the table with assignment specifies the desired next y state of a par-

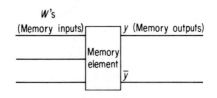

Figure 12-1

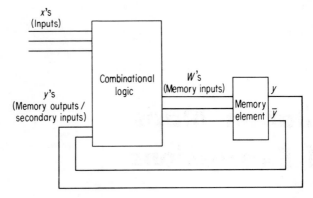

Figure 12-2

ticular memory element. For any memory element, the present y state of the element, the desired next y state, and the characteristics of that type of element specify the W states, or memory excitations, required to produce the desired next y state; this relationship is dependent on the type of element, and is independent of any table requirements. The desired next y state requirement is thus translated into a present W state requirement. The combination of the memory element relationships with the specific table requirements results in the present x and y states specifying present W states. Recall that the "W" symbols have been used for generality; they will now be replaced with specific symbols used for the various memory element inputs.

Memory elements for pulse operation will be discussed first. They fall in the category called flip-flops. A *flip-flop* is a memory element having two stable states. A flip-flop may have one, two, or three inputs, and has two complementary level outputs, y and $\bar{y}$. Recall that the memory element outputs reflect the state of the memory element; when $y = 1$, $\bar{y} = 0$, the flip-flop is said to be "on"; when $y = 0$, $\bar{y} = 1$, the flip-flop is "off."

Six flip-flops will be studied: *S-C*, *J-K*, *T*, *S-C-T*, *J-K-T*, and *D*. The name of each flip-flop corresponds to its input names. The symbols S, C, J, K, T, and D thus represent the memory element inputs and replace the general W.

The standard symbols place the input and output labels inside the block; they use the labels 1 and 0 for the flip-flop outputs, rather than y and $\bar{y}$. A flip-flop can be thought of as a logic device that stores a single bit of information; the output, 1 or 0, that is "on" signifies the bit stored. This nomenclature has the obvious drawback that either output, 1 or 0, can be in the 1- or 0-state, and this could lead, to confusion. One must remain aware that the 1 and 0 are simply output labels and do not indicate the output states. We use the standard in the ensuing figures, Figs. 12-3 through 12-8, but retain the y and $\bar{y}$ on the output lines.

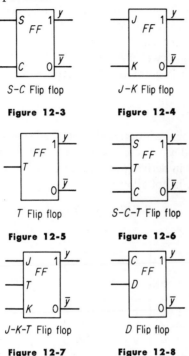

S–C Flip flop   J–K Flip flop

Figure 12-3   **Figure 12-4**

T Flip flop   S–C–T Flip flop

Figure 12-5   **Figure 12-6**

J–K–T Flip flop   D Flip flop

Figure 12-7   **Figure 12-8**

The six flip-flops will be discussed in the order mentioned. Two methods for obtaining memory element excitation expressions for pulse operation will be given.

Excitation Expressions in Pulse Operation

S-C Flip-flop

The S-C flip-flop (Fig. 12-3) has two inputs, S (set) and C (clear). If the flip-flop is off, a signal on the S input will turn it on; a signal on the C input will cause no change. If the flip-flop is on, a signal on the C input will turn it off; a signal on the S input will cause no change. The S and C inputs of this flip-flop must never be signaled simultaneously, since the resulting state is indeterminate.

This flip-flop is also commonly known as the *S-R* (set-reset) flip-flop. Although "*S-R*" is to be preferred over "*S-C*" because of the possibility of confusion between *C* (clear) and *C* (clock), we submit to the standard symbol.

SC				
y	00	01	11	10
0	0	0	–	1
1	1	0	–	1

S–C Characteristic map

Figure 12-9

The *S-C* characteristics can be summarized in a *characteristic map* (Fig. 12-9). The map variables are the flip-flop inputs, *S* and *C*, and the present state of the flip flop, *y*. Each map entry specifies the next state of the flip-flop; an optional entry signifies a nonallowable input combination.

We shall use *y'* to represent the next state of a memory element. (Note: *y'* is *not* being used to represent the complement of *y*.) Reading the map gives us the *characteristic equation* of the flip-flop:

$$y' = S + \bar{C}y$$

Using this equation, we can derive the required states of *S* and *C* for each combination of *y* and *y'*. These are summarized in the following table. It must be kept in mind in deriving the values of *S* and *C* to satisfy the equations that $S = C = 1$ is not allowed.

S-C State Requirements

y	y'	$y' = S + \bar{C}y$	S	C
0	1	$1 = S + \bar{C} \cdot 0$	1	0
1	0	$0 = S + \bar{C} \cdot 1$	0	1
1	1	$1 = S + \bar{C} \cdot 1$	–	0
0	0	$0 = S + \bar{C} \cdot 0$	0	–
Optional			–	–

The equation method was given for the interested reader, The required *S* and *C* states can more easily be derived directly from the characteristic map, as shown in (Figs. 12-10 through 12-13).

The *S-C* state requirements can be described in words as follows. To turn the flip-flop on ($y = 0$, $y' = 1$), there must be a signal on the *S* input ($S = 1$) and there must be no signal on the *C* input ($C = 0$). To turn the flip-flop off ($y = 1$, $y' = 0$), there must be no signal on the *S* input ($S = 0$) but there must be a signal on the *C* input ($C = 1$). To keep the flip-flop on ($y = 1$, $y' = 1$), there may or may not be a signal on the *S* input ($S = $ —) and there must be no signal on the *C* input ($C = 0$). To keep the flip-flop off ($y = 0$, $y' = 0$), there must be no signal on the *S* input ($S = 0$) and there may or may not be a signal on the *C* input ($C = $ —).

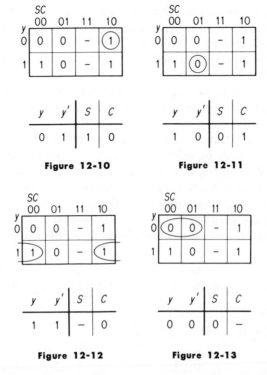

Figure 12-10

Figure 12-11

Figure 12-12   Figure 12-13

An optional entry may appear in the state table because a certain combination of present memory element states can never occur, or because for a certain combination of present memory element states, a particular input signal can never occur; or because for a certain combination of present memory element states, we don't care what the circuit action is for a particular input signal. For optional table entries, all input states for any type of memory element are optional.

The *S-C* state requirement information is now combined with any specific table requirements to obtain *S-C* flip-flop excitation maps; from these maps, memory excitation expressions are read which implement the table requirements with *S-C* flip-flops. The state table with assignment from Fig. 11-2 will be used as an example, and is repeated here (Fig. 12-14).

$y_1 y_2$	x_1	x_2
00	11/0	00/0
11	01/0	00/0
01	10/0	00/1
10	10/0	00/0

Figure 12-14

A memory element excitation map is drawn for each combination of circuit pulse input, or clocked input combination, and memory element input. For example, if there are two circuit pulse inputs, x_1 and x_2, and three *S-C* flip-flops required, twelve maps are drawn. A convenient arrangement is to have all maps corresponding to a particular memory element input in the same row, and all maps corresponding to a particular circuit pulse input, or clocked input combination, in the same column.

The outputs of all memory elements are variables for all maps. In addition, the circuit pulse input, or clocked input combination, defining a map is a variable for that map; remember that these variables are mutually exclusive—hence the separate maps.

In the present example (Fig. 12-14), two S-C flip-flops will be used; with the two circuit pulse inputs, eight memory element excitation maps are drawn (Fig. 12-15).

The required S and C states for each combination of y and y' can be entered in these maps by reference to the table with assignment. Consider the first row of the table, in which the present flip-flop state is $y_1y_2 = 00$. If an x_1 pulse occurs, the next flip-flop state is $y'_1y'_2 = 11$; the states $S = 1$ and $C = 0$ are therefore required for both flip-flops (Fig. 12-15).

If an x_2 pulse occurs when $y_1y_2 = 00$, the next state is $y'_1y'_2 = 00$; the states $S = 0$ and $C = -$ are required for both flip-flops (Fig. 12-16).

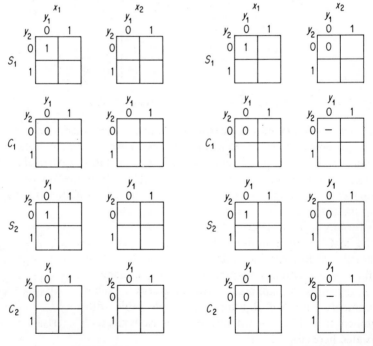

Figure 12-15   **Figure 12-16**

Now consider the second row of the table, in which the present flip-flop state is $y_1y_2 = 11$. If an x_1 pulse occurs, $y'_1y'_2 = 01$; the states $S = 0$ and $C = 1$ are required for flip-flop 1; $S = -$ and $C = 0$ are required for flip-flop 2 (Fig. 12-17).

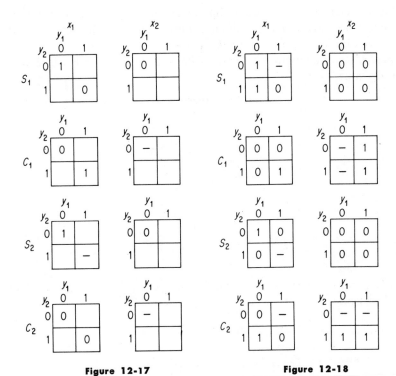

Figure 12-17   **Figure 12-18**

The rest of the entries are made in the same manner; the completed memory element excitation maps are shown in Fig. 12-18.

The *S-C* flip-flop excitation expressions are now read from the maps:

$$S_1 = x_1 \bar{y}_1$$
$$C_1 = x_1 y_1 y_2 + x_2$$
$$S_2 = x_1 \bar{y}_1 \bar{y}_2$$
$$C_2 = x_1 \bar{y}_1 y_2 + x_2$$

The characteristics of the other memory elements will be discussed and applied in a similar manner.

J-K Flip-flop

The *J-K* flip-flop (Fig. 12-4) has two inputs, *J* and *K*, and has the same characteristics as the *S-C* flip-flop (the *J* input corresponds to the *S* input, and the *K* input corresponds to the *C* input) with one exception: both inputs may be pulsed simultaneously, in which case the

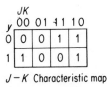

J — K Characteristic map

Figure 12-19

flip-flop will change state. The *J-K* flip-flop should be clocked to ensure that both inputs are indeed pulsed simultaneously. The *J-K* characteristics are summarized in the map in Fig. 12-19.

The *J-K* state requirements derived from the map are summarized in the following table:

J-K State Requirements

y y'	J	K
0 1	1	—
1 0	—	1
1 1	—	0
0 0	0	—
Optional	—	—

J-K flip-flops applied to the table in the previous example (Fig. 12-14) yield the memory excitation maps shown in Fig. 12-20.

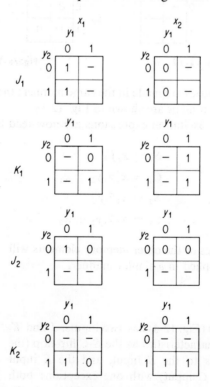

Figure 12-20

The *J-K* flip-flop excitation expressions are

$$J_1 = x_1$$
$$K_1 = x_1 y_2 + x_2$$
$$J_2 = x_1 \bar{y}_1$$
$$K_2 = x_1 \bar{y}_1 + x_2$$

T Flip-flop

The *T* flip-flop (Fig. 12-5) has one input, *T* (trigger).[1] If the flip-flop is off, a pulse on the *T* input will turn it on. If the flip-flop is on, a pulse on the *T* input will turn it off. A pulse on the input thus always changes the flip-flop state. The *T* characteristics are summarized in the map in Fig. 12-21.

T Characteristic map

Figure 12-21

The *T* state requirements derived from the map are summarized in the following table:

T State Requirements

y	y'	T
0	1	1
1	0	1
1	1	0
0	0	0
Optional		—

T flip-flops applied to the previous example yield the memory element excitation maps shown in Fig. 12-22.

The *T* flip-flop excitation expressions are

$$T_1 = x_1 \bar{y}_1 + x_1 y_2 + x_2 y_1$$
$$T_2 = x_1 \bar{y}_1 + x_2 y_2$$

Figure 12-22

Note that the common term $x_1 \bar{y}_1$ need be implemented only once.

S-C-T Flip-flop

The *S-C-T* flip-flop (Fig. 12-6) has three inputs *S*, *C*, and *T*, and has the combined characteristics of the *S-C* flip-flop and the *T* flip-flop. If the flip-

---

[1] Also called toggle.

flop is off, a pulse on the S input or T input or both will turn it on; a pulse on the C input will cause no change. If the flip-flop is on, a pulse on the C input or T input or both will turn it off; a pulse on the S input will cause no change.

The S and C inputs of this flip-flop must never be pulsed simultaneously, since the resulting state is indeterminate. For the same reason, the S and T inputs must never be pulsed simultaneously when the flip-flop is on, and the C and T inputs must never be pulsed simultaneously when the flip-flop is off.

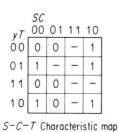

S−C−T Characteristic map

Figure 12-23

The S-C-T characteristics are summarized in the map in Fig. 12-23.

The S-C-T state requirements derived from the map are summarized in the following table:

S-C-T State Requirements

y y'	S	C	T
0 1	1	0	—
	—		1
1 0	0	1	—
		—	1
1 1	—	0	0
0 0	0	—	0
Optional	—	—	—

The S-C-T state requirements for turning on the flip-flop can be described in words as follows: there must be a pulse on the S input or T input; the other may or may not be pulsed; there must not be a pulse on the C input.

The state requirements for turning off the flip-flop can be described analogously.

S-C-T flip-flops applied to the previous example yield the excitation maps shown in Fig. 12-24.

These maps must be examined concurrently because of the built-in OR characteristic of the S-C-T flip-flop; the flip-flop can be turned on by a pulse on the S or T input, and it can be turned off by a pulse on the C or T input. Every map entry with a 1 also includes a —, and each such entry appears in two maps, S and T or C and T. This signifies that at least one of

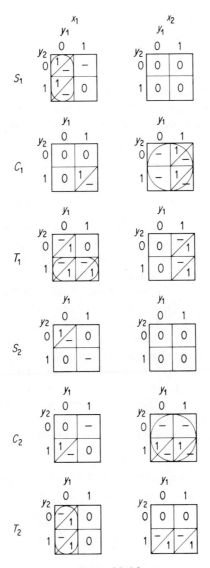

Figure 12-24

the two entries must be accounted for; the other may be treated as optional. The selected groups read are indicated on the maps.

The *S-C-T* flip-flop excitation expressions are

$$S_1 = x_1 \bar{y}_1$$

$$C_1 = x_2$$

$$T_1 = x_1 y_2$$
$$S_2 = \text{(unused)}$$
$$C_2 = x_2$$
$$T_2 = x_1 \bar{y}_1$$

$S_1 = x_1 \bar{y}_2$ could have been selected; however, the term $x_1 \bar{y}_1$ was chosen since it is common to both S_1 and T_2 and need be implemented only once. Note that the groups selected eliminate the need for an OR, which would be necessary with the solutions

$S_1 = \text{(unused)}$	$S_1 = x_1 \bar{y}_1$
$C_1 = x_2$ or	$C_1 = x_1 y_1 y_2 + x_2$
$T_1 = x_1 \bar{y}_1 + x_1 y_2$	$T_1 = \text{(unused)}$

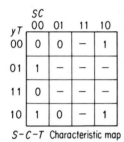

$S{-}C{-}T$ Characteristic map
No simultaneous changes

Figure 12-25

If the S and T inputs are to be pulsed simultaneously when the flip-flop is off, or the C and T inputs are to be pulsed simultaneously when the flip-flop is on, the $S{-}C{-}T$ flip-flop should be clocked; otherwise, if the T pulse lags the S or C pulse, a double change in state could occur.

In some of the literature, the $S{-}C{-}T$ flip-flop is *arbitrarily* restricted to no two inputs ever being pulsed simultaneously, the resulting state being said to be indeterminate. This might be the case, however, only for an unclocked flip-flop.

The $S{-}C{-}T$ characteristics for this version are summarized in the map in Fig. 12-25.

The $S{-}C{-}T$ state requirements derived from the map are summarized in the following table:

S-C-T State Requirements—No Simultaneous Changes

y y'	S	C	T
0 1	1	0	0
	0		1
1 0	1	0	0
	0		1
1 1	—	0	0
0 0	0	—	0
Optional	—	—	—

The effect of this restriction is that the number of optional entries in the excitation maps is reduced. Every former entry with a 1 and — is replaced with a 1 and 0; this signifies that exactly one of each pair of corresponding entries must be accounted for; the other may *not* be treated as optional. The first definition of the *S-C-T* flip-flop is to be preferred since it leads to more optional entries and therefore generally more economical excitation expressions.

The only effect on the previous example is that $S_1 = x_1 \bar{y}_2$ must be selected since the entry $x_1 \bar{y}_1 y_2$ is accounted for by T_1, and cannot be accounted for twice.

J-K-T Flip-flop

The *J-K-T* flip-flop (Fig. 12-7) has three inputs, *J, K,* and *T,* and has the same characteristics as the *S-C-T* flip-flop with one exception: the *J* and *K* inputs may be pulsed simultaneously, in which case the flip-flop will change state. The *J-K-T* flip-flop thus has the combined characteristics of the *J-K* flip-flop and the *T* flip-flop.

When the flip-flop is on, the *J* and *T* inputs must never be pulsed simultaneously unless the *K* input is also pulsed. When the flip-flop is off, the *K* and *T* inputs must never be pulsed simultaneously unless the *J* input is also pulsed.

The *J-K-T* characteristics are summarized in the map in Fig. 12-26.

yT	JK 00	01	11	10
00	0	0	1	1
01	1	—	1	1
11	0	0	0	—
10	1	0	0	1

J-K-T Characteristic map

Figure 12-26

The *J-K-T* state requirements derived from the map are summarized in the following table:

J-K-T State Requirements

y y′	J	K	T
0 1	1	—	—
	—	0	1
1 0	—	1	—
	0	—	1
1 1	—	0	0
0 0	0	—	0
Optional	—	—	—

The *J-K-T* state requirements for turning on the flip-flop can be described in words as follows: there must be a pulse on the *J* input or *T* input; the other may or may not be pulsed; there may be a pulse on the *K* input only if there is a pulse on the *J* input.

The state requirements for turning off the flip-flop can be described analogously.

The reader may want to verify that the *J-K-T* flip-flop excitation expressions for the previous example are the same as those for the *S-C-T* flip-flop.

D Flip-flop

The *D* flip-flop (Fig. 12-8) has one input, *D* (delay), plus a clock, *C*. It is also commonly called the *D* element or *clocked delay element*. The *D* flip-flop is most generally used in clocked pulse operation, with the *D* input being a level input. If the *D* input is on at the time of the clock pulse, the flip-flop will turn on, or remain on; if the *D* input is off at the time of the clock pulse, the flip-flop will turn off, or remain off. An example of this circuit action is illustrated in Fig. 12-27.

In practice, the *D* flip-flop may employ a longer "clock pulse"—in reality, a clock level—and "latch" on the leading edge or trailing edge of the clock; variations of the basic element have such names as *edge-sensitive D* or *polarity-hold latch*.

The *D* characteristics are summarized in the map in Fig. 12-28.

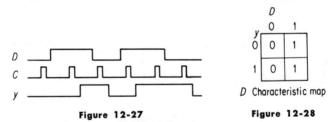

Figure 12-27

D Characteristic map

Figure 12-28

The *D* state requirements derived from the map are summarized in the following table:

D State Requirements

y	y'	D
0	1	1
1	0	0
1	1	1
0	0	0
Optional		—

Note that y' is a function solely of D; the present excitation will become the next state.

D flip-flops applied to the previous example yield the memory element excitation maps shown in Fig. 12-29.

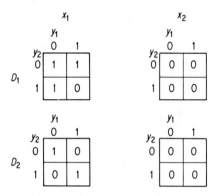

Figure 12-29

The D flip-flop excitation expressions are

$$D_1 = x_1 \bar{y}_1 + x_1 \bar{y}_2$$
$$D_2 = x_1 \bar{y}_1 \bar{y}_2 + x_1 y_1 y_2$$

Universal Map Method

The conventional method just described for obtaining pulse operation excitation expressions can be represented schematically as in Fig. 12-30. For each combination of pulse input, or clocked input combination, and memory element required, a memory element excitation map is drawn for each input of each type of element considered. The memory element state requirements define the rules for how each map is drawn from the table. Each map is read for the excitation expression.

A method will now be described which can be represented schematically as in Fig. 12-31. In this method, for each combination of pulse input, or clocked input combination, and memory element required, a single memory element excitation map is drawn. This map is a function only of the table, and is completely independent of any type of memory element that may be considered. The memory element state requirements define the rules for how this universal map is read to obtain the excitation expressions for each type of memory element input.

The universal map requires five symbols, one for each possible combination of y and y', plus the optional symbol. The symbols used are shown in

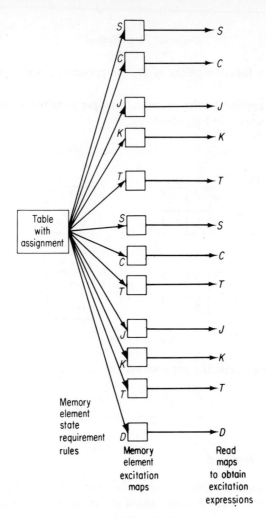

S → S

C → C

J → J

K → K

T → T

S → S

C → C

T → T

J → J

K → K

T → T

D → D

Table with assignment

Memory element state requirement rules

Memory element excitation maps

Read maps to obtain excitation expressions

Figure 12-30

the following table:

y	y'	Map Symbol
0	1	**1**
1	0	**0**
1	1	1
0	0	0
Optional		—

A summary of the state requirements for the six memory elements discussed is given in Fig. 12-32; the map symbols are also shown in the figure,

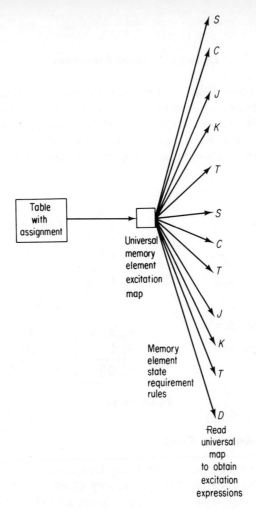

Figure 12-31

Table with assignment → Universal memory element excitation map

Memory element state requirement rules

Read universal map to obtain excitation expressions

y y'	Map symbol	S	C	J	K	T	S	C	T	J	K	T	D
0 1	**1**	1	0	1	—	1	1	0	1 / —	— / 1	1 / 0	— / 1	1
1 0	**0**	0	1	—	1	1	1	0	1 / —	— / 1	— / 0	1 / —	0
1 1	1	—	0	—	0	0	—	0	0	—	0	0	1
0 0	0	0	—	0	—	0	0	—	0	0	—	0	0
Optional	—	—	—	—	—	—	—	—	—	—	—	—	—

Summary of memory element state requirements

Figure 12-32

and their relationship to each of the state requirements, in terms of how the map is read to obtain excitation expressions, will now be discussed.

The table in Fig. 12-14 will again be used for an example, and is repeated in Fig. 12-33, along with the universal maps from the table. Note that the table and the maps contain the same excitation information but in different forms.

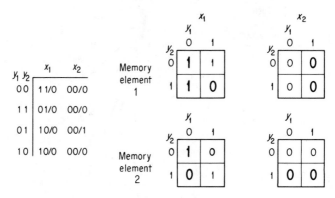

Figure 12-33

Universal Map-Reading Rules

The rules for reading the universal map for each memory element input can be derived from the table of the memory element state requirements. Basically, in each memory element input column, a 1 indicates that the map symbol in the corresponding row must be accounted for in the excitation expression for that input; a — indicates that the map symbol in the corresponding row may be used optionally in the excitation expression.

A portion of the table of the memory element state requirements is reproduced for an example:

Map Symbol	S
1	1
0	0
1	—
0	0
—	—

Examining the S column, the 1 in the first row indicates that the map symbol **1** must be accounted for in the S excitation expression; the — in the third and fifth rows indicate that the map symbols 1 and — may be used optionally in the S excitation expression.

A modification occurs with memory elements with the built-in OR characteristic, e.g., the S-C-T or J-K-T flip-flops, in which there is a choice of accounting for a map symbol in either of two excitation expressions; this modification will be apparent when the S-C-T and J-K-T flip-flops are discussed.

While the rules are here formally derived from the table of the memory element state requirements, they can also be easily realized intuitively simply by the consideration of the characteristics of each particular memory element.

S-C Flip-flop

The map-reading rules for the S-C flip-flop (see Fig. 12-32) are

Every **1** must be accounted for in the S expression.
Every **0** must be accounted for in the C expression.
Any 1 or — may be used optionally in the S expression.
Any 0 or — may be used optionally in the C expression.

From the maps in Fig. 12-33, the S-C flip-flop excitation expressions are

$$S_1 = x_1 \bar{y}_1$$
$$C_1 = x_1 y_1 y_2 + x_2$$
$$S_2 = x_1 \bar{y}_1 \bar{y}_2$$
$$C_2 = x_1 \bar{y}_1 y_2 + x_2$$

The reader can compare the results with the conventional method and the universal map method and note that both methods, as should be expected, produce the same excitation expressions for each memory element.

J-K Flip-flop

The map-reading rules for the J-K flip-flop (see Fig. 12-32) are

Every **1** must be accounted for in the J expression.
Every **0** must be accounted for in the K expression.
Any **0**, 1, or — may be used optionally in the J expression.
Any **1**, 0, or — may be used optionally in the K expression.

From the maps in Fig. 12-33, the *J-K* flip-flop excitation expressions are

$$J_1 = x_1$$
$$K_1 = x_1 y_2 + x_2$$
$$J_2 = x_1 \bar{y}_1$$
$$K_2 = x_1 \bar{y}_1 + x_2$$

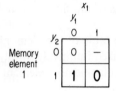

Figure 12-34

Another example of interest is given in Fig.12-34; only a pertinent portion of the problem is shown. Reading this map for a *J-K* flip-flop,

$$J_1 = x_1 y_2$$
$$K_1 = x_1$$

T Flip-flop

The map-reading rules for the *T* flip-flop (see Fig. 12-32) are

Every **1** and **0** must be accounted for in the *T* expression.
Any — may be used optionally in the *T* expression.

From the maps in Fig. 12-33, the *T* flip-flop excitation expressions are

$$T_1 = x_1 \bar{y}_1 + x_1 y_2 + x_2 y_1$$
$$T_2 = x_1 \bar{y}_1 + x_2 y_2$$

S-C-T Flip-flop

A portion of the table of the memory element state requirements is reproduced:

Map Symbol	S	C	T
1	1	0	—
	—		1

Note that a choice is available: The map symbol **1** must be accounted for either in the *S* expression or the *T* expression. A **1** accounted for in the *S* expression may be used optionally in the *T* expression, and vice versa. A choice is also available with the map symbol **0**.

The map-reading rules for the *S-C-T* flip-flop (see Fig. 12-32) are

Every **1** must be accounted for either in the *S* expression or the *T* expression.

Every **0** must be accounted for either in the *C* expression or the *T* expression.

Any **1** accounted for in the *T* expression, or any 1 or — may be used optionally in the *S* expression.

Any **0** accounted for in the *T* expression, or any 0 or — may be used optionally in the *C* expression.

Any **1** accounted for in the *S* expression, any **0** accounted for in the *C* expression, or any — may be used optionally in the *T* expression.

From the maps in Fig. 12-33, the *S-C-T* flip-flop excitation expressions are

$$S_1 = x_1 \bar{y}_1$$
$$C_1 = x_2$$
$$T_1 = x_1 y_2$$
$$S_2 = \text{(unused)}$$
$$C_2 = x_2$$
$$T_2 = x_1 \bar{y}_1$$

A portion of another example is given in Fig. 12-35. The point of interest in this example centers on some possible variations in the *C* and *T* expressions; only these expressions will be considered:

$$\begin{matrix} C_1 = x_1 y_1 + x_2 \\ T_1 = \text{(unused)} \end{matrix} \quad \text{or} \quad \begin{matrix} C_1 = x_1 y_1 \\ T_1 = x_2 y_2 \end{matrix} \quad \text{or} \quad \begin{matrix} C_1 = x_2 \\ T_1 = x_1 y_1 y_2 \end{matrix}$$

The first solution requires two logic blocks (an AND and an OR); the second solution also requires two logic blocks (two AND's); the third solution re-

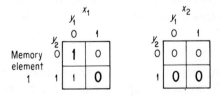

Figure 12-35

quires only one logic block (an AND). The last solution illustrates that it may sometimes be desirable to use larger terms in order to take optimum advantage of the built-in OR characteristic of a memory element.

J-K-T Flip-flop

The map-reading rules for the *J-K-T* flip-flop (see Fig. 12-32) are

Every **1** must be accounted for either in the *J* expression or the *T* expression.

Every **0** must be accounted for either in the *K* expression or the *T* expression.

Any **1** accounted for in the *T* expression, any **0** accounted for in the *K* expression, or any 1 or — may be used optionally in the *J* expression.

Any **0** accounted for in the *T* expression, any **1** accounted for in the *J* expression, or any 0 or — may be used optionally in the *K* expression.

Any **1** accounted for in the *J* expression, any **0** accounted for in the *K* expression, or any — may be used optionally in the *T* expression.

From the maps in Fig. 12-33, the *J-K-T* flip-flop excitation expressions are the same as those for the *S-C-T* flip-flop.

D Flip-flop

The map-reading rules for the *D* flip-flop (see Fig. 12-32) are

Every **1** and 1 must be accounted for in the *D* expression.
Any — may be used optionally in the *D* expression.

From the maps in Fig. 12-33, the *D* flip-flop excitation expressions are

$$D_1 = x_1 \bar{y}_1 + x_1 \bar{y}_2$$
$$D_2 = x_1 \bar{y}_1 \bar{y}_2 + x_1 y_1 y_2$$

The following example is used to review the reading of the universal map for the six memory elements discussed.

EXAMPLE

A pulse operation sequential circuit with one pulse input requires four memory elements. A universal map for one of these memory elements is shown in Fig. 12-36.

The excitation expressions read from the map are

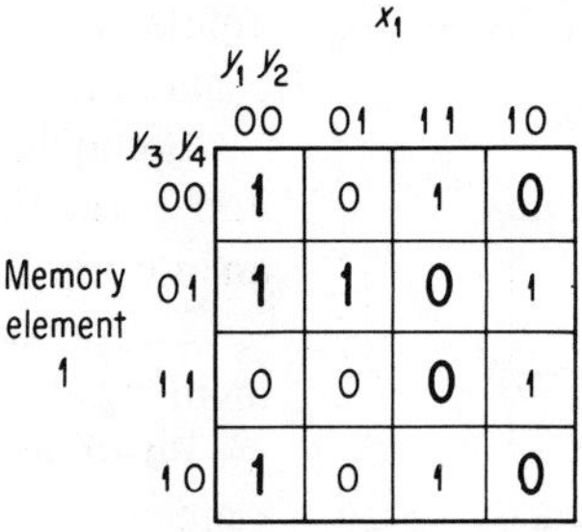

$$S_1 = x_1\bar{y}_1\bar{y}_2\bar{y}_4 + x_1\bar{y}_1\bar{y}_3y_4$$

$$C_1 = x_1y_1\bar{y}_2\bar{y}_4 + x_1y_1y_2y_4$$

$$J_1 = x_1\bar{y}_2\bar{y}_4 + x_1\bar{y}_3y_4$$

$$K_1 = x_1\bar{y}_2\bar{y}_4 + x_1y_2y_4$$

$$T_1 = x_1\bar{y}_2\bar{y}_4 + x_1\bar{y}_1\bar{y}_3y_4 + x_1y_1y_2y_4$$

$$S_1 = x_1\bar{y}_1\bar{y}_3y_4$$

$$C_1 = x_1y_1y_2y_4$$

$$T_1 = x_1\bar{y}_2\bar{y}_4$$

$$J_1 = x_1\bar{y}_3y_4$$

$$K_1 = x_1y_2y_4$$

$$T_1 = x_1\bar{y}_2\bar{y}_4$$

$$D_1 = x_1\bar{y}_1\bar{y}_2\bar{y}_4 + x_1\bar{y}_1\bar{y}_3y_4 + x_1y_1y_2\bar{y}_4 + x_1y_1\bar{y}_2y_4$$

Figure 12-36

Note that the "rules" in both the conventional and universal map methods are the same; they are the memory element state requirements. The difference lies in where in the procedure the rules are applied. In the conventional method, they are applied in constructing the maps; in the universal map method, they are applied in reading the single map.

In the universal map method, there is, of course, the obvious advantage that only one map need be drawn; the excitation expressions for all inputs of *any* type of memory element, those discussed here or others, can be read from this one map. Another advantage is that with memory elements in which the inputs must be examined concurrently, such as the *S-C-T* or *J-K-T* flip-flops, the optimum expressions are more easily determined from the single map. Still another advantage is that the optimum choice of memory element is often obvious from inspection of the single map, and it is not necessary to go through the actual derivation and comparison of many memory element excitation expressions.

Miscellany

The circuit cost is a function not only of the combinational logic implementing the excitation expressions, but also of the type of memory element used, since the types of memory elements may differ in cost.

The same type of memory element does not have to be used throughout a circuit; for example, memory element 1 could be an *S-C* flip-flop, memory element 2 a *T* flip-flop, etc.

Some flip-flops have a special clock input (e.g., the *D* flip-flop or see Fig. 12-37), the flip-flops having the equivalent of an internal AND for a clock pulse and level input combinations.

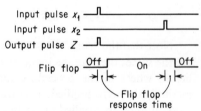

To achieve the desired "power on" state, an initial memory element reset may be required.

Positive or negative logic may be used with memory element inputs and outputs, as with combinational logic blocks.

Figure 12-37

A few reminders about timing relationships: An input pulse to a memory element may be switched with an output on this same element, the element entering into its own control. See, for example, Fig. 12-38.

The relative durations of input pulses and flip-flop response time may be depicted by the timing chart in Fig. 12-39.

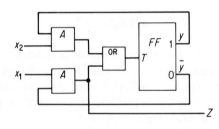

Figure 12-38

Figure 12-39

When the flip-flop is off ($\bar{y} = 1$), an x_1 input pulse, switched with the $\bar{y}$ output, turns the flip-flop on. When the flip-flop is on ($y = 1$), an x_2 input pulse, switched with the y output, turns the flip-flop off.

Note, by reference to the circuit diagram and timing chart, that the output pulse Z is coincident with an x_1 input pulse occurring when the flip-flop is off, even though this same pulse initiates the turning on of the flip-flop.

Because of the relative durations of input pulses and memory element response time, two or more memory elements may simultaneously be caused to change state by the same input pulse without a concern about race conditions. See, for example, Fig. 12-40.

When both flip-flops are off, an x_1 input pulse, switched with the $\bar{y}_1$ and $\bar{y}_2$ outputs, turns both flip-flops on. Even if one flip-flop responds faster than the other, no race condition exists. For example, if it is assumed that flip-flop 1 responds faster than flip-flop 2, the circuit action may be depicted by the timing chart in Fig. 12-41.

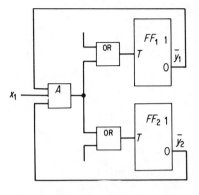

Figure 12-40

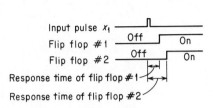

Figure 12-41

Excitation Expressions in Level Operation

S-C Flip-flop

The *S-C* flip-flop is the only memory element so far discussed that can be used with level operation. The other five memory elements either have a "triggering" characteristic or require a clock pulse, both of which are compatible only with pulse operation.

The table with assignment in Fig. 12-42 will be used as an example. For completeness, both the conventional and universal map methods will be shown.

$y_1 y_2$ \ $x_1 x_2$	00	01	11	10
00	00	01	--	10
01	11	01	11	--
11	11	10	11	11
10	00	10	11	10

Figure 12-42

With the conventional method, a map is drawn for each input of the two memory elements required (Fig. 12-43).

The *S-C* flip-flop excitation expressions are

$$S_1 = x_1 + \bar{x}_2 y_2$$
$$C_1 = \bar{x}_1 \bar{x}_2 \bar{y}_2$$
$$S_2 = x_1 x_2 + x_2 \bar{y}_1$$
$$C_2 = \bar{x}_1 x_2 y_1$$

With the universal map method, a map is drawn for each of the two memory elements required (Fig. 12-44).

The map-reading rules for the *S-C* flip-flop will produce the same *S-C* flip-flop excitation expressions as just obtained.

Y Element

The simplest memory element of all is the *Y* element, or *delay element* (Fig. 12-45). With a delay (memory element transition time) of Δ, the signal

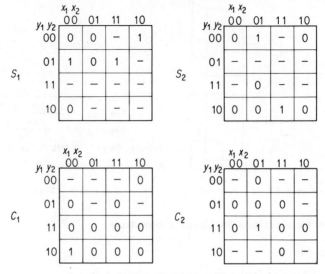

Figure 12-43

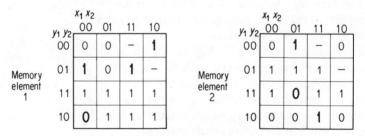

Figure 12-44

on the input, Y, at time t, appears on the output, y, at time $t + \Delta$. y is the present state, and Y is the present excitation. The next state, y', will be the same as the present excitation, Y.

The other memory elements discussed are relatively long-term memories; the Y element is a relatively short-term memory. An example of its circuit action is illustrated in Fig. 12-46.

The delay is usually inherent because of the signal propagation time around the feedback loop, or a device may be inserted. All feedback loops,

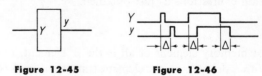

Figure 12-45   **Figure 12-46**

as a practical matter, must have gain greater than unity, so that the circuits involved are self-sustaining; if inherent amplification is not already present, a device may be inserted for this purpose.

Theoretically, the Y element could be used in pulse operation if the pulse or clock inputs could be synchronized with the y pulses. This timing requirement is virtually impossible to realize, however, and the Y element is practical with level operation only.

The Y characteristics are summarized in the map in Fig. 12-47.

The Y-state requirements derived from the map are summarized in the following table:

	Y	
y	0	1
0	0	1
1	0	1

Y Characteristic map

Figure 12-47

Y-State Requirements

y y'	Y
0 1	1
1 0	0
1 1	1
0 0	0
Optional	—

Note that the D flip-flop and Y element have the same characteristics and state requirements; the D flip-flop, or clocked delay element, is applicable to pulse operation, while the Y element or delay element, is applicable to level operation.

The Y element excitation expressions for the previous example can be read from the maps in Fig. 12-44. The map-reading rules for the Y element are the same as those for the D flip-flop:

Every **1** and 1 must be accounted for in the Y expression.
Any — may be used optionally in the Y expression.

However, the Y expressions, because $Y = y'$, can be read directly from the table with assignment. In Fig. 12-42, all left-hand entries define Y_1 ($= y'_1$), and all right-hand entries define Y_2 ($= y'_2$).

If the Y_1 and Y_2 entries were mapped separately, the Y maps in Fig. 12-48 would result. (These maps should be compared with those in Fig. 12-44.) One should, however, practice reading the Y excitation expressions directly from the table with assignment. If only Y elements are to be con-

$$
\begin{array}{c|cccc}
 & \multicolumn{4}{c}{x_1 x_2} \\
y_1 y_2 & 00 & 01 & 11 & 10 \\
\hline
00 & 0 & 0 & - & 1 \\
01 & 1 & 0 & 1 & - \\
11 & 1 & 1 & 1 & 1 \\
10 & 0 & 1 & 1 & 1 \\
\end{array}
\qquad
\begin{array}{c|cccc}
 & \multicolumn{4}{c}{x_1 x_2} \\
y_1 y_2 & 00 & 01 & 11 & 10 \\
\hline
00 & 0 & 1 & - & 0 \\
01 & 1 & 1 & 1 & - \\
11 & 1 & 0 & 1 & 1 \\
10 & 0 & 0 & 1 & 0 \\
\end{array}
$$

Y_1 map Y_2 map

Figure 12-48

sidered, there is no need to draw any maps at all; in level operation, the tables have essentially been transformed into Y maps.

In the example, the Y excitation expressions are

$$Y_1 = x_1 + \bar{x}_2 y_2 + x_2 y_1$$
$$Y_2 = x_1 x_2 + x_2 \bar{y}_1 + \bar{x}_2 y_2$$

PROBLEMS

1. From the table with assignment in Fig. 12-49, obtain the universal excitation maps.

*2. From the table with assignment in Fig. 12-50, obtain the universal excitation maps.

$$
\begin{array}{c|cc|c}
 & x_1 & x_2 & Z \\
y_1 y_2 & & & \\
\hline
00 & 10 & 00 & 0 \\
10 & 01 & 11 & 1 \\
01 & 01 & 00 & 0 \\
11 & 10 & 11 & 1 \\
\end{array}
\qquad
\begin{array}{c|cc}
 & x_1 & x_2 \\
y_1 y_2 & & \\
\hline
00 & 00/0 & 01/0 \\
11 & 11/0 & 01/1 \\
01 & 11/0 & 10/0 \\
10 & 00/0 & 10/0 \\
\end{array}
$$

Figure 12-49   **Figure 12-50**

3. Obtain the memory element excitation expressions indicated in Fig. 12-51.

*4. A set-dominant S-C flip-flop has the same characteristics as the S-C flip-flop with one exception: both inputs may be pulsed simultaneously, in which case the S input dominates and the flip-flop will turn on or remain on. Derive the universal map-reading rules for this flip-flop.

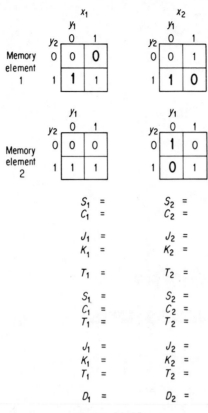

$$S_1 =$$
$$C_1 =$$

$$S_2 =$$
$$C_2 =$$

$$J_1 =$$
$$K_1 =$$

$$J_2 =$$
$$K_2 =$$

$$T_1 =$$

$$T_2 =$$

$$S_1 =$$
$$C_1 =$$
$$T_1 =$$

$$S_2 =$$
$$C_2 =$$
$$T_2 =$$

$$J_1 =$$
$$K_1 =$$
$$T_1 =$$

$$J_2 =$$
$$K_2 =$$
$$T_2 =$$

$$D_1 =$$

$$D_2 =$$

Figure 12-51

5. Obtain the Y excitation expressions from the table with assignment in Fig. 12-52. Find all minimum sum of products and product of sums solutions.

*6. Obtain the Y excitation expressions from the table with assignment in Fig. 12-53. Find all minimum sum of products and product of sums solutions.

$y_1 y_2$ \ $x_1 x_2$	00	01	11	10
00	00	00	10	01
01	11	01	01	01
11	11	10	11	11
10	--	00	11	--

Figure 12-52

$y_1 y_2$ \ $x_1 x_2$	00	01	11	10
00	00	01	10	00
01	11	01	01	01
11	11	11	11	10
10	--	--	11	00

Figure 12-53

213

13

Output Maps
and Expressions

Having derived the excitation expressions, it now remains to obtain the output, or Z expressions, and the combinational logic portion of the sequential circuit will be completed and the design can be implemented.

In this chapter we shall discuss obtaining the Z expressions and the problem of hazards; and, finally, the entire method of sequential circuit synthesis will be summarized with a few examples complete from word statement to final circuit diagram.

Output Expressions in Pulse Operation

The output expressions in pulse operation can be read directly from the table with assignment. The rows of the table may be reordered so that the memory element states are in reflected ordering.

A few examples are given in Figs. 13-1 and 13-2. Note that the output expressions are independent of the type of memory element used.

Output Expressions in Level Operation

Obtaining the output expressions in level operation is a more involved procedure than in pulse operation. Recall that the primitive flow table and the

$y_1 y_2$	x_1	x_2
00	00/0	01/0
01	10/0	11/1
11	00/0	11/0
10	10/0	01/1

$Z = x_2 \bar{y}_1 y_2 + x_2 y_1 \bar{y}_2$

(Pulse output)

Figure 13-1

$y_1 y_2$	x_1	x_2	Z
00	00	01	0
01	10	11	1
11	00	11	0
10	10	01	1

$Z = \bar{y}_1 y_2 + y_1 \bar{y}_2$

(Level-output)

Figure 13-2

flow table with assignment were retained; from these two tables, an output, or Z, map is constructed, and the output expressions are read from the map.

The output state for each stable state is identified in the primitive flow table; the location, in the Z map, of the output state is identified in the flow table with assignment. Also, the actual state-to-state transitions are identified in the primitive flow table, this information being used in the assignment of output states for the unstable states.

Figure 13-3 shows a primitive flow table and a corresponding flow table with assignment.

In constructing the Z map, the output state for each stable state is entered in the map first. The output state for each stable state is found in the primitive flow table; the location, in the Z map, of the output state corresponds to the location of the associated stable state in the flow table with assignment. The partially completed Z map is shown in Fig. 13-4.

Partially Completed Z Map

$x_1 x_2$: 00	01	11	10	Z
(1)	2	3	4	0
-	(2)	5	4	0
1	2	(3)	4	1
1	2	5	(4)	1
-	2	(5)	4	0

Primitive flow table

y \ $x_1 x_2$	00	01	11	10
0	(1)	2	(3)	4
1	1	(2)	(5)	(4)

Flow table with assignment

Figure 13-3

y \ $x_1 x_2$	00	01	11	10
0	0			1
1		0	0	1

Figure 13-4

In the assignment of output states for the unstable states, the following rules are observed:

1. If, in a transition, the states of an output for the initial and final stable states are the same, this same output state must be assigned for all unstable states involved in the transition. Transient changes in output state are thus prevented.

2. The output states for all unstable states not covered by rule 1 may be optional except that in all transitions involving a change in output state, the output must change state only once. The exception must be noted when there are two or more unstable states involved in a transition; oscillatory changes of output states are thus prevented.

The preceding Z map will now be completed. The flow table with assignment indicates that unstable state 2 may be involved in a transition from stable state ① to ② or from ③ to ②. The ① to ② transition specifies no change in output state (the initial output state is 0, and the final output state is 0). If this transition does in fact occur, the output state must be 0 for unstable state 2. Reference to the primitive flow table shows that the ① to ② transition does occur; therefore, the output state 0 must be assigned to unstable state 2.

The flow table with assignment indicates that unstable state 4 may be involved in a transition from ① to ④ or from ③ to ④. The ③ to ④ transition is the critical one in this case; both the initial and final output states are 1. Reference to the primitive flow table shows that the ③ to ④ transition occurs; therefore, the output state must be 1 for unstable state 4.

The flow table with assignment indicates that unstable state 1 may be involved in a transition from ② to ①, from ⑤ to ①, or from ④ to ①. The ② to ① and ⑤ to ① transitions both specify no change in output state; if *either* transition occurs, the output state must be 0 for unstable state 1. The primitive flow table shows, however, that neither of these transtions occurs. Since unstable state 1 is not involved in a transition specifying no change in output state, the output state may be optional for unstable state 1. Figure 13-5 shows the completed Z map and the output expressions.

Figure 13-6 shows an example with two outputs. The primitive flow table is not shown, but from it has been obtained the output state for each stable state; these output states appear in the partially completed Z map. All left-hand map entries define Z_1; all right-hand map entries define Z_2. The primitive flow table also shows that all transitions indicated in the flow table with assignment do occur.

$x_1 x_2$

y	00	01	11	10
0	0	0	1	1
1	−	0	0	1

$$Z = x_1 \bar{y} + x_1 \bar{x}_2$$
$$\text{or } x_1 \bar{y} + \bar{x}_2 y$$

Figure 13-5

$x_1 x_2$

$y_1 y_2$	00	01	11	10
00	①	2	③	4
01	1	②	5	④
11	6	7	⑤	8
10	⑥	⑦	3	⑧

Flow table
with assignment

$x_1 x_2$

$y_1 y_2$	00	01	11	10
00	00		11	
01		01		10
11			01	
10	10 00			11

Partially completed
Z map

Figure 13-6

The Z map will now be completed. Unstable state 2 is involved in transitions ① to ②, output states $00 \rightarrow 01$; and ③ to ②, output states $11 \rightarrow 01$. Examination of the two outputs independently shows that the ① to ② transi-

	Output States	
Transition	Z_1	Z_2
① to ②	$0 \rightarrow 0$	$0 \rightarrow 1$
③ to ②	$1 \rightarrow 0$	$1 \rightarrow 1$

tion specifies no change in the Z_1 output state, requiring $Z_1 = 0$ for unstable state 2; and that the ③ to ② transition specifies no change in the Z_2 output state, requiring $Z_2 = 1$ for unstable state 2. Therefore, $Z_1 Z_2 = 01$ is required for unstable state 2.

The output state requirement, $Z_1 Z_2 = 10$, for unstable state 4 is similarly determined: The ① to ④ transition requires $Z_2 = 0$, and the ③ to ④ transition requires $Z_1 = 1$. Unstable state 1 requires the output state $Z_1 Z_2 = 00$, the ② to ① transition requiring $Z_1 = 0$, and the ④ to ① transition requiring $Z_2 = 0$. Unstable state 5 requires the output state $Z_1 Z_2 = 01$, the ② to ⑤ transition establishing both the Z_1 and Z_2 output states. The output states for unstable state 6 can be optional, since in the ⑤ to ⑥ transition, both outputs change state. The ⑤ to ⑦ transition requires the output state $Z_1 = 0$ for unstable state 7, whereas Z_2 can be optional since it change state. The ⑤ to ⑧ transition requires the output state $Z_2 = 1$ for unstable state 8 whereas Z_1 can be optional since it changes state. Unstable state 3 requires the output state $Z_1 Z_2 = 11$.

The completed Z map and the output expressions are shown in Fig. 13-7.

There might be a requirement that if a transition involves a multiple output change, all output state changes are to occur simultaneously. For example, with this requirement, in the ⑤ to ⑥ transition in the preceding example, the output states for unstable state 6 would be restricted to either $Z_1 Z_2 = 01$ or $Z_1 Z_2 = 10$.

	$x_1 x_2$			
$y_1 y_2$	00	01	11	10
00	00	01	11	10
01	00	01	01	10
11	--	0-	01	-1
10	10	00	11	11

$Z_1 = x_1 \bar{y}_2 + x_1 \bar{x}_2 + \bar{x}_2 y_1$
$Z_2 = x_2 \bar{y}_1 + x_1 y_1$

Figure 13-7

Timing considerations may sometimes take precedence over circuit economy, and definite output states may be assigned in place of optional ones. For example,

If it is desired that all outputs be of as short a duration as possible, all optional entries can be replaced by 0's,

If it is desired that all outputs be of as long a duration as possible, all optional entries can be replaced by 1's,

If it is desired that all output changes occur as soon as possible, all optional entries can be replaced by the output entries for the corresponding final stable states,

If it is desired that all output changes occur as late as possible, all optional entries can be replaced by the output entries for the corresponding initial stable states.

Such timing considerations may, of course, apply to particular transtions only.

If the outputs are sampled only when stable, the output states for all unstable states may be optional.

Some examples of output state assignments for unstable states involved in cycles and races follow.

Cycle—No Change in Output State

See Fig. 13-8. $Z = 0$ must be assigned to all unstable states so that no transient output change occurs.

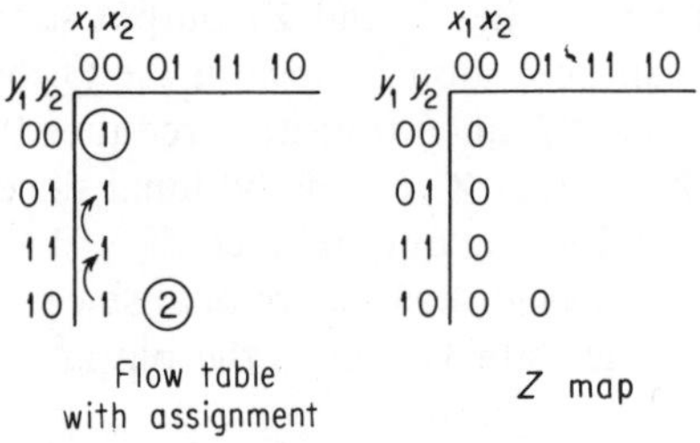

Flow table
with assignment

Z map

Figure 13-8

Cycle—Change in Output State

See Fig. 13-9. The output must change state only once, and therefore the choice of optional output states is restricted to the solutions in Fig. 13-10.

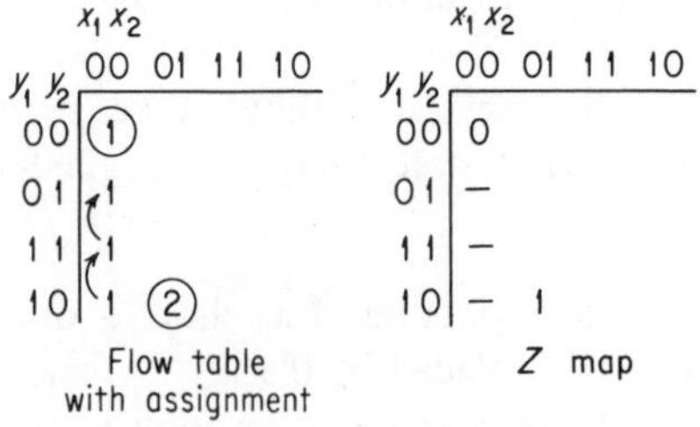

Flow table
with assignment

Z map

Figure 13-9

$y_1 y_2$ \ $x_1 x_2$	00	01	11	10
00	0			
01	0			
11	0			
10	0			1

$y_1 y_2$ \ $x_1 x_2$	00	01	11	10
00	0			
01	0			
11	0			
10	1			1

$y_1 y_2$ \ $x_1 x_2$	00	01	11	10
00	0			
01	0			
11	1			
10	1			1

$y_1 y_2$ \ $x_1 x_2$	00	01	11	10
00	0			
01	1			
11	1			
10	1			1

Z maps

Figure 13-10

Race—No Change in Output State

See Fig. 13-11. $Z = 0$ must be assigned to all unstable states so that no transient output change occurs.

$y_1 y_2$ \ $x_1 x_2$	00	01	11	10
00	①			
01	1			
11	1	②		
10	1			

Flow table with assignment

$y_1 y_2$ \ $x_1 x_2$	00	01	11	10
00	0			
01	0			
11	0	0		
10	0			

Z map

Figure 13-11

Race—Change in Output State

See Fig. 13-12. The output must change state only once, and therefore the choice of optional output states is restricted to the solutions in Fig. 13-13.

$y_1 y_2$ \ $x_1 x_2$	00	01	11	10
00	①			
01	1			
11	1	②		
10	1			

Flow table with assignment

$y_1 y_2$ \ $x_1 x_2$	00	01	11	10
00	0			
01	—			
11	—	1		
10	—			

Z map

Figure 13-12

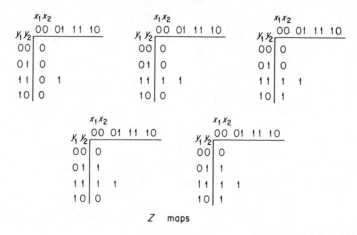

Z maps

Figure 13-13

Transient Outputs; Cyclic Specifications

Transient outputs, associated only with particular *unstable* states, may sometimes be specified in a sequential circuit requirement. For example, in the flow table in Fig. 13-14, a transient output associated with unstable state 2 might be desired, the expression for this output being

$$Z = \bar{x}_1 x_2 \bar{y}_1 \bar{y}_2$$

Cycles may be prescribed for the express purpose of introducing a series of transient outputs, as in the example in Fig. 13-15.

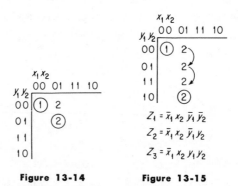

Figure 13-14   **Figure 13-15**

A continuous series of transient outputs is sometimes desired, as in the example in Fig. 13-16. When an input change to $x_1 x_2 = 01$ occurs, the circuit will cycle continuously, producing the series of transient outputs until another input change occurs.

When a cycle is required as part of the original circuit specifications, such as in furnishing a series of transient outputs, transitions *from* and *to* the same rows of a flow table occur in the same column, imposing restrictions on the applicability of the assignment patterns previously discussed. For example, in the left three columns of the flow table in Fig. 13-17, transitions between all six pairs of rows occur. As far as these left three columns are concerned, the assignment (pattern #5, Chapter 11) in Fig. 13-18 is satisfactory. However, this assignment is not satisfactory for the $x_1x_2 = 10$ column. If the *a* to *b* transition is prescribed by the cycle *afgb*, the *b* to *c* transition must be prescribed by the cycle *bhc*, and there are then no spare rows adjacent to *c* to accomplish the *c* to *d* transition. If the *a* to *b* transition is prescribed by the alternative cycle *aehb*, the *b* to *c* transition must be prescribed by the cycle *bgc*, and again there are no spare rows adjacent to *c* to accomplish the *c* to *d* transition.

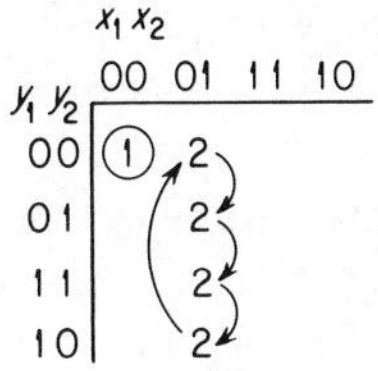

$Z_1 = \bar{x}_1 x_2 \bar{y}_1 \bar{y}_2$
$Z_2 = \bar{x}_1 x_2 \bar{y}_1 y_2$
$Z_3 = \bar{x}_1 x_2 y_1 y_2$
$Z_4 = \bar{x}_1 x_2 y_1 \bar{y}_2$

Figure 13-16

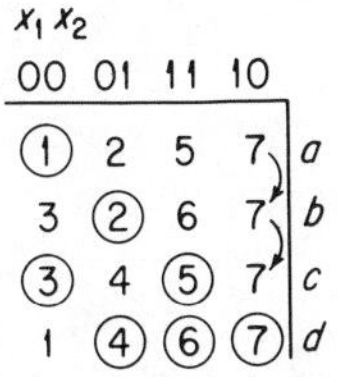

Figure 13-17

y_3 \ y_1y_2	00	01	11	10
0	*a*	*c*	*h*	*e*
1	*f*	*g*	*b*	*d*

Figure 13-18

Alternative assignments with the same pattern may be applicable when such a condition exists. For instance, the assignment in Fig. 13-19 is satisfactory.

Pattern #6, Chapter 11, is unique in that it is applicable to any four-row flow table, regardless of any type of cycle that may be prescribed.

y_3 \ y_1y_2	00	01	11	10
0	*a*	*b*	*h*	*e*
1	*f*	*g*	*c*	*d*

Figure 13-19

Hazards

The physical devices used to implement switching circuits are not ideal in the sense that the relationships

$$\text{if} \quad X = 0, \quad \text{then} \quad \bar{X} = 1$$
$$\text{if} \quad X = 1, \quad \text{then} \quad \bar{X} = 0$$

do not always exist. During transition times, the relationships

$$X = \bar{X} = 0$$
$$\text{or} \quad X = \bar{X} = 1$$

may briefly exist. The consequences, in sequential circuits, of this imperfection in devices will now be studied.

In the circuit in Fig. 13-20, if there were an inherent delay in the implementation of the NOT logic block, the timing relationship shown in Fig. 13-21 would result.

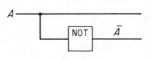

Figure 13-20   **Figure 13-21**

In the implementation of

$$AB + \bar{A}C$$

or its equivalent

$$(A + C)(\bar{A} + B)$$

this delay would produce the spurious signals shown in Fig. 13-22. Such false signals, caused by a momentary $X = \bar{X}$ relationship, are called *hazards*.

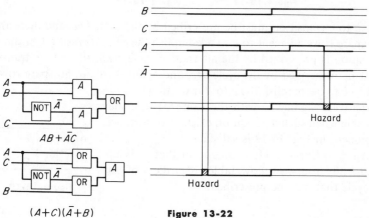

Figure 13-22

Hazards are generally undesirable because of the unwanted "glitch" in the signal. They become a very serious problem, however, if the false signal is presented to a memory element input in level operation, in which case the circuit might enter an incorrect state. Hazards must therefore be checked for and eliminated.

Maps can be used in the identification and elimination of possible hazards. The map associated with the preceding example is shown in Fig. 13-23.

AB

C	00	01	11	10
0	0	0	1	0
1	1	1	1	0

Figure 13-23

The expression $AB + \bar{A}C$ is obtained by grouping the 1-squares as in Fig. 13-24. A hazard can exist when a circuit change causes a movement between two states not in the same group, for example, between $ABC = 011$ and 111, as indicated by the arrows in Fig. 13-24. The hazard can be eliminated by grouping the 1-squares between which this movement exists, as shown in the map and corresponding circuit in Fig. 13-25. The hazard is eliminated since the logic block corresponding to the term BC maintains the circuit output in the *on* state when $BC = 11$.

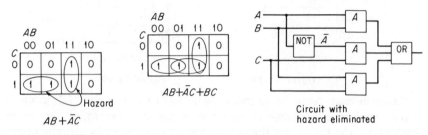

Figure 13-24

Figure 13-25

The expression $(A + C)(\bar{A} + B)$ is obtained by using the complementary approach and grouping the 0-squares on the map as in Fig. 13-26. A hazard can exist in this case, when there is a circuit change between $ABC = 000$ and 100, as indicated by the arrows. The hazard can be eliminated by grouping these 0-squares, as shown in the map and corresponding circuit in Fig. 13-27. The hazard is eliminated since the logic block corresponding to the term $(B + C)$ maintains the circuit output in the *off* state when $BC = 00$.

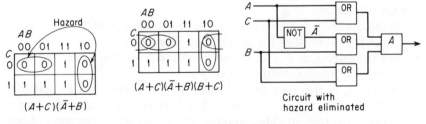

Figure 13-26

Figure 13-27

To eliminate hazards in sequential circuits, redundancy is thus sometimes required.

Another example is shown in Fig. 13-28. If a circuit is implemented from the expression $\bar{C}\bar{D} + B\bar{C} + AD$, a hazard can exist when there is a circuit change between $ABCD = 1000$ and 1001, as indicated by the arrows. The hazard can be eliminated by adding the term $A\bar{C}$ to the expression, and implementing $\bar{C}\bar{D} + B\bar{C} + AD + A\bar{C}$. Note that the prime implicant $A\bar{C}$ would be added, rather than the term $A\bar{B}\bar{C}$, which is not a prime implicant.

If the circuit is implemented from the expression $(A + B + \bar{D})(A + \bar{C})$ $(\bar{C} + D)$, obtained by the complementary approach, Fig. 13-29, no hazard can exist.

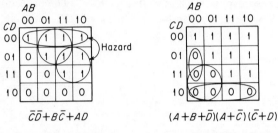

Figure 13-28   Figure 13-29

Hazards may occur in memory excitation (Y) or output (Z) circuits. Before modifying a circuit to eliminate a possible hazard, it should be determined whether or not the corresponding condition can actually occur. This is done by reference to the flow table with assignment, the primitive flow table, and the physical implementation. The flow table with assignment may show that the hazard cannot exist because the associated circuit change can never occur. On the other hand, if the flow table indicates that the change may occur, the change is verified in the primitive flow table. The primitive flow table may show that the circuit change can never occur and that therefore the hazard cannot exist. However, if it is verified that the change can occur, the type of change—0 to 1, 1 to 0, or both—and the states of the other variables are correlated with the physical implementation (see, for example, Fig. 13-22) for the final determination of whether the condition can actually occur.

If the condition can never occur, then no hazard actually exists. If the condition can occur, then the hazard must be eliminated.

The variables A, B, C, and D, used in this section, were chosen for generality, and may represent either x's or y's.

A hazard, in the most general sense, is an incorrect circuit operation due to delays in the physical devices. The hazards discussed in this section are called static hazards. The interested reader can find other types of hazards, e.g., dynamic and essential hazards, discussed in the literature.

Illustrative Problem—Pulse Operation

A pulse operation sequential circuit is to have two pulse inputs, x_1 and x_2, and one pulse output, Z. The output pulse Z is to be coincident with the first x_2 pulse immediately following exactly two consecutive x_1 pulses. (See Figs. 13-30 through 13-33.)

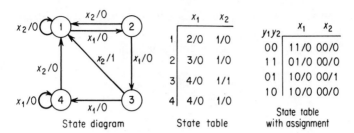

State diagram State table State table with assignment

Figure 13-30 **Figure 13-31**

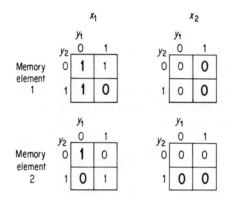

Memory element excitation maps

Figure 13-32

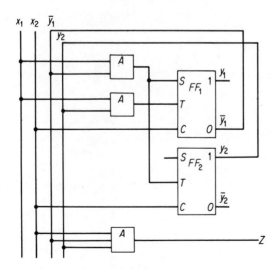

Circuit diagram

Figure 13-33

The state table in Fig. 13-30 cannot be reduced, and the assignment in Fig. 13-31 is arbitrarily chosen.

S-C-T flip-flops are selected for implementation:

$$S_1 = x_1 \bar{y}_1$$
$$C_1 = x_2$$
$$T_1 = x_1 y_2$$
$$S_2 = \text{(unused)}$$
$$C_2 = x_2$$
$$T_2 = x_1 \bar{y}_1$$
$$Z = x_2 \bar{y}_1 y_2$$

As in the case of any multi-output circuit, the entire circuit should be evaluated as a whole, since the various excitation and output circuits may be able to share logic blocks in common.

Illustrative Problem—Clocked Pulse Operation

This problem should be compared with the previous one. A clocked pulse operation sequential circuit is to have two level inputs, x_1 and x_2, and one clock, C. An output pulse, Z, is to be coincident with a clock pulse occurring with $x_1 x_2 = 01$ immediately following exactly two clock pulses with $x_1 x_2 = 10$. $x_1 x_2 = 00$ and $x_1 x_2 = 11$ can never occur. (See Figs. 13-34 through 13-37.)

With S-C-T flip-flops,

$$S_1 = C x_1 \bar{x}_2 \bar{y}_1$$
$$C_1 = C \bar{x}_1 x_2$$
$$T_1 = C x_1 \bar{x}_2 y_2$$
$$S_2 = \text{(unused)}$$
$$C_2 = C \bar{x}_1 x_2$$
$$T_2 = C x_1 \bar{x}_2 \bar{y}_1$$
$$Z = C \bar{x}_1 x_2 \bar{y}_1 y_2$$

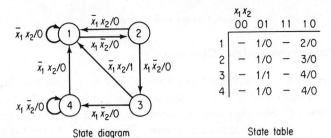

State diagram   State table

Figure 13-34

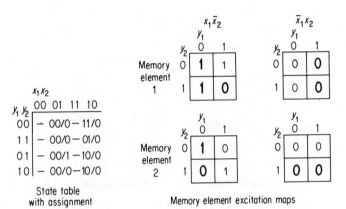

$y_1 y_2$	00	01	11	10
00	–	00/0	–	11/0
11	–	00/0	–	01/0
01	–	00/1	–	10/0
10	–	00/0	–	10/0

State table
with assignment

Figure 13-35

Memory element excitation maps

Figure 13-36

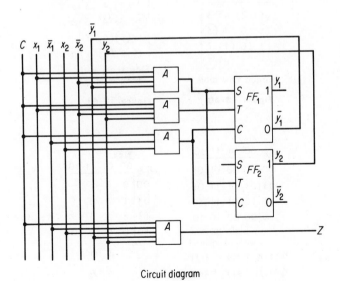

Circuit diagram

Figure 13-37

Illustrative Problem—Level Operation

A level operation sequential switching circuit is to have two inputs, x_1 and x_2, and one output, Z. Z is to turn on when x_1 turns on; Z is to turn off when x_2 turns off. No other input sequence is to cause any change in output. Only one input can change state at a time. The circuit is to be implemented with Y elements. (See Figs. 13-38 through 13-44.)

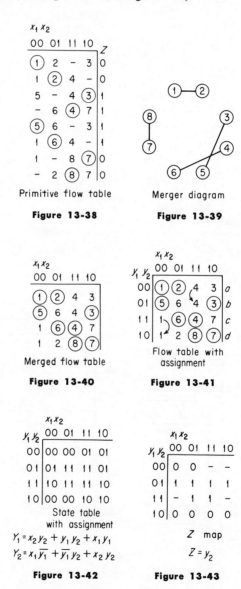

Primitive flow table

Figure 13-38

Merger diagram

Figure 13-39

Merged flow table

Figure 13-40

Flow table with assignment

Figure 13-41

State table with assignment

$$Y_1 = x_2 y_2 + y_1 y_2 + x_1 y_1$$
$$Y_2 = x_1 \overline{y_1} + \overline{y_1} y_2 + x_2 y_2$$

Figure 13-42

Z map

$$Z = y_2$$

Figure 13-43

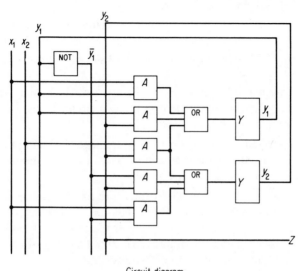

Circuit diagram

Figure 13-44

PROBLEMS

1. From the flow tables in Fig. 13-45, draw the Z map and obtain the output expressions.

Figure 13-45

*2. Using another merger of the primitive flow table in Fig. 13-45, draw the Z map and obtain the output expressions.

3. Design a sequential circuit for the requirements in Example §2, Chapter 9,
   (a) Using assignment #1 on page 162.
   (b) Using assignment #2 on page 162.

(c) Using assignment #3 on page 162.

A solution requiring two two-input AND circuits as the total combinational circuit requirement is possible with assignments #1 and #2, and a solution requiring one three-input AND circuit is possible with assignment #3.

4. Design a sequential circuit for the requirements in Example §3, Chapter 9. A solution requiring two two-input AND circuit as the total combinational circuit requirement is possible.

*5. Design a sequential circuit for the requirements in Example §5, Chapter 9. A solution requiring one two-input AND circuit as the total combinational circuit requirement is possible.

*6. Analyze the circuit in Fig. 13-46 and write a word statement describing the sequential circuit action. From the word statement, design a more economical circuit. An assignment other than the one used in the circuit in Fig. 13-46 leads to a solution requiring only two two-input AND circuits as the total combinational circuit requirement.

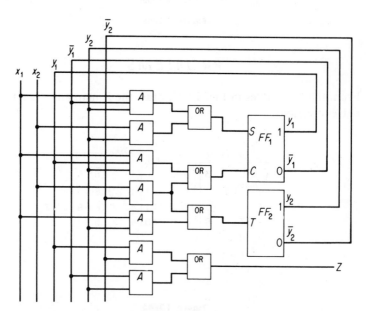

Figure 13-46

7. A level input sequential circuit is to have two inputs, x_1 and x_2 and one output, Z. The output is to be on only when $x_1 x_2 = 01$, or when $x_1 x_2 = 11$ immediately following $x_1 x_2 = 01$. The output state is optional for $x_1 x_2 = 11$ immediately following $x_1 x_2 = 00$. All input changes are possible. The circuit is to be implemented with Y elements.

*8. A level input sequential circuit is to have two inputs, x_1 and x_2, and one output Z. Z is to turn on when x_2 turns on, provided that x_1 is on at the time. Z is to turn off when x_2 turns off. If x_1 changes state simultaneously with x_2 turning on, the circuit action is optional. All input changes are possible. The circuit is to be implemented with Y elements.

Appendices

Relay Circuits

A

Contact Networks

The relationship of Boolean expressions and contact networks will be examined. Contacts may be operated by several means such as switches, keys, cams, or relays. The discussion will be in terms of relays, which can operate a number of contacts simultaneously. The implementation of simpler devices, such as switch contacts, will be obvious.

Relay contact terminology includes the terms *springs* and *contacts*. There

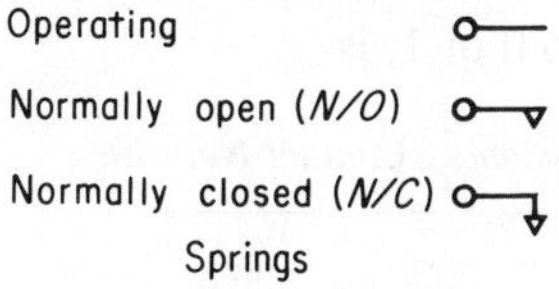

Operating   o—

Normally open (*N/O*)   o—▽

Normally closed (*N/C*) o—⌐↓

Springs

Figure A-1

N/O
Contacts

Figure A-2

are three types of springs as shown pictorially in Fig. A-1. The discussion will be limited, for the time being, to two types of contacts: normally open, *N/O*, made up of two springs (Fig. A-2), and normally closed, *N/C*, made up of two springs (Fig. A-3). (*N/O* contacts are also called "make" contacts; *N/C* contacts are also called "break" contacts.)

N/C
Contacts

Figure A-3

235

A relay consists of a coil (electromagnet) and a set of contacts. When current is passed through the coil, all of the operating springs are caused to move from their normal, or unoperated, position to their operated position. With current through the coil, the relay is said to be *operated*; with no current through the coil, the relay is said to be *unoperated*, or in its normal state.[1] Thus, normally open contacts are open when the relay is unoperated, and closed when the relay is operated. Normally closed contacts are closed when the relay is unoperated, and open when the relay is operated. By convention, contacts are drawn in their normal state (eg., as in Figs. A-2 and A-3).

Relay coil and contact relationships can be represented pictorially as shown in Fig. A-4.

The general schematic diagram for a relay contact network can be represented as shown in Fig. A-5.

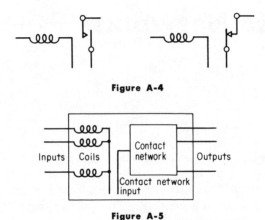

Figure A-4

Figure A-5

Contact networks, including individual contacts, can be in one of two possible states: *open* or *closed*. The relationship to the corresponding Boolean expression, which may be equal to 0 or 1, is

Boolean Expression		Contact Network
0	=	Open
1	=	Closed

In a Boolean expression, the variables, which may equal 0 or 1, relate to the relays, which may be unoperated or operated, as follows:

Boolean Variable		Relay
0	=	Unoperated
1	=	Operated

---

[1]Relay transition time is neglected here.

For example, the Boolean expression $A + \bar{B}C$ relates to a relay contact network as follows: $A + \bar{B}C = 1$ if $A = 1$, or if $B = 0$ and $C = 1$. The related contact network is closed if relay A is operated, or if relay B is unoperated and relay C is operated.

Suppose the following simple circuit requirement: a circuit is to be closed if relay X is operated. The Boolean expression for the circuit requirement is simply

$$X$$

and the circuit would be drawn pictorially as in Fig. A-6. When relay X is operated, the N/O contact is closed.

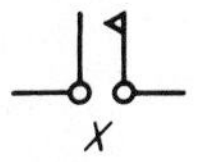

Suppose now, another, equally simple circuit requirement: a circuit is to be closed if relay X is unoperated. The Boolean expression for this circuit is

Figure A-6

$$\bar{X}$$

and the circuit would be drawn pictorially as in Fig. A-7. When relay X is unoperated, the N/O contact is closed.

In the preceding two examples, note the relationship between an uncomplemented literal and its corresponding N/O contact, and the relationship between a complemented literal and its corresponding N/C contact. This relationship gives the convenient symbolic notation for relay contacts shown in Fig. A-8. Each uncomplemented literal, in a Boolean expression relating to a contact network, corresponds to a N/O contact; each complemented literal in the expression corresponds to a N/C contact.

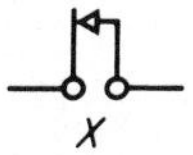

$N/O\ X$ contact

$N/C\ X$ contact

Figure A-7 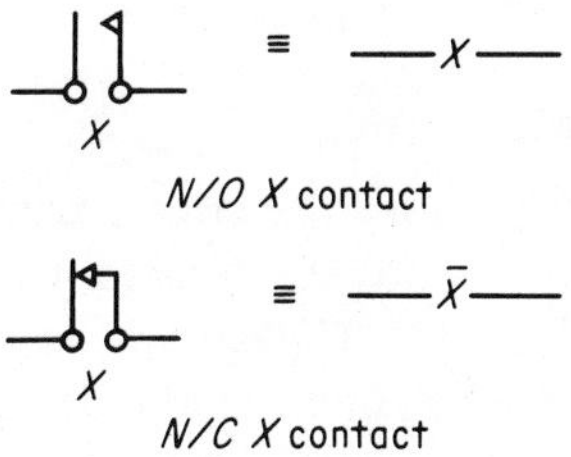   **Figure A-8**

Implementation of AND, OR, and NOT Functions

AND

Suppose that a relay contact network is to be closed only if relays A AND B are operated. The Boolean expression for this circuit is

$$AB$$

and the network requires a N/O contact on relay A and a N/O contact on relay B. For the network to be closed only when both relays A AND B are operated, these contacts must be placed in series (Fig. A-9). Thus, the Boolean AND function is realized in contact networks by a series connection.

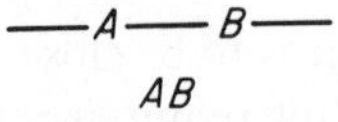

—A——B—

AB

Figure A-9 $\cdot$ = AND = series connection

OR

Suppose that a contact network is to be closed if relay A OR B is operated. The Boolean expression for this circuit is

$$A + B$$

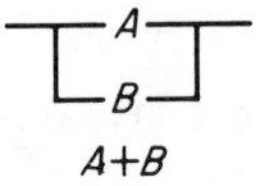

Again, a N/O contact is required on each relay. For the network to be closed if relay A OR B is operated, a parallel connection is required (Fig. A-10). Thus, the Boolean OR function is realized in contact networks by a parallel connection.

$A+B$

Figure A-10

$+$ = OR = parallel connection

NOT

The NOT function, as previously shown, is implemented by the use of normally closed contacts. Thus, if a relay contact network is to be closed if relay A is NOT operated, the Boolean expression would be

$$\bar{A}$$

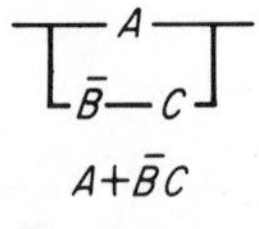

and the circuit would be realized by the use of a normally closed contact on A.

$A+\bar{B}C$

The circuit to realize the function $A + \bar{B}C$, discussed earlier, would be as in Fig. A-11. It should be noted that a Boolean expression is related to a single-input single-output (two-terminal) *series-parallel* contact network.

Figure A-11

The examples in Fig. A-12 show the application of a few Boolean algebra theorems to the simplification of contact networks. In the first example, the normally closed A contact is redundant. If the B contact is closed, the network will be closed regardless of the state of relay A; if relay A is unoperated, the $\bar{A}B$ path closes the network; if relay A is operated, the A path closes the network. The second example illustrates the application of the included-term theorem. The BC path is redundant since there is also a B

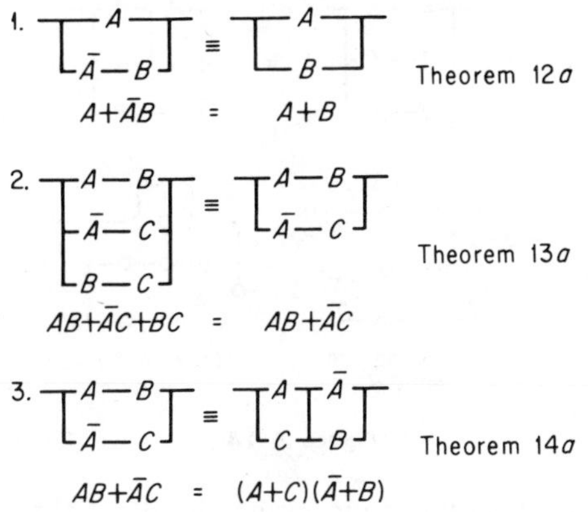

1. $A + \bar{A}B \;=\; A + B$ Theorem 12*a*

2. $AB + \bar{A}C + BC \;=\; AB + \bar{A}C$ Theorem 13*a*

3. $AB + \bar{A}C \;=\; (A+C)(\bar{A}+B)$ Theorem 14*a*

Figure A-12

contact in the AB path and a C contact in the AC path. If both of these contacts B and C are closed, then the network will be closed because relay A must be in one state or the other, and either the normally open A contact or the normally closed A contact must be closed. In the third example, the transposition theorem is applied to the resultant circuit from the second example. Note that the transposition reintroduces the included path BC.

We shall now go beyond the simplification of Boolean functions and examine further simplifications peculiar to contact networks: transfer contacts, bridge circuits, nonplanar networks, and graphical complementation.

Transfer Contacts

Transfer contacts are made up of a normally open contact and a normally closed contact sharing a common operating spring. Thus, they are made up of three spring (Fig. A-13). Transfer contacts in which the operating spring opens one contact before closing the other contact are called "break-before-make" transfer contacts. With these transfer contacts, there is a brief period of time during relay operation when both the normally open and normally closed contacts are open. Transfer contacts in which the operating spring closes one contact before opening the other contact are called "make-before-break" or "continuity-transfer" contacts. With these transfer contacts, there is a brief period of time during relay operation when both the normally open and normally closed contacts are closed.

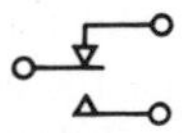

Figure A-13

In a contact network, it is desirable to bring together in an optimum manner normally open and normally closed contacts, on the same relay, to make transfer contacts. A normally open and a normally closed contact

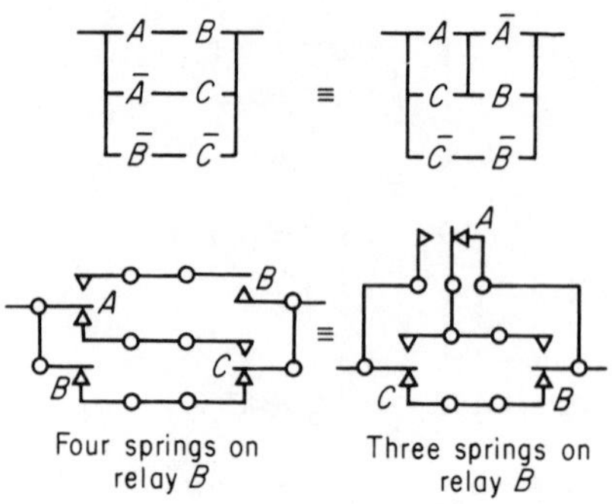

$$\begin{bmatrix} A - B \\ \bar{A} - C \\ \bar{B} - \bar{C} \end{bmatrix} \equiv \begin{bmatrix} A - \bar{A} \\ C - B \\ \bar{C} - \bar{B} \end{bmatrix}$$

Four springs on   Three springs on
   relay B relay B

Figure A-14

that are not combined require four springs in all; if they are combined to make a transfer contact, only three springs are required. An example of this is shown in Fig. A-14.

Sometimes relays are built up of springs as needed, and it is desirable to minimize the number of springs. In other cases, relays come standardly equipped with a fixed number of transfer contacts. For instance, a particular type of relay might be available in three sizes: 4, 6, or 12 transfer contacts; these are referred to as 4-, 6-, or 12-position relays. In each position a transfer contact is available. If a circuit specifies a transfer contact, one position on the relay is required. If a circuit specifies a N/O contact, again one position is required, the normally closed spring being left unused. If a circuit specifies a N/C contact, one position is required, the normally open spring being left unused. Thus, a N/O contact and a N/C contact that are not combined require two relay positions in all; if they can be combined into a transfer contact, only one position is required.

Each literal in a Boolean expression corresponds to a contact in the associated series-parallel network. Therefore, the minimization of the Boolean expression leads to a minimum series-parallel contact requirement. Optimizing the number of transfer contacts enables us to minimize the number of springs or positions, whichever is the criterion. The minimization of positions is especially important if it leads to a smaller standard relay being required, since the cost of a relay is a function of its size. For instance, if seven positions on a relay are needed, a 12-position relay could be required. However, if the circuit can be redesigned to require only six positions, then a 6-position relay could suffice.

If, in a relay contact network,

$$P = \text{the total number of positions}$$
$$S = \text{the total number of springs}$$
and
$$C = \text{the total number of contacts}$$

the following relationship exists:

$$P = S - C$$

The following table (which includes Figs. A-15, A-16, and A-17) summarizes some of the relationships discussed:

	N/O Contact	N/C Contact	Transfer Contacts
	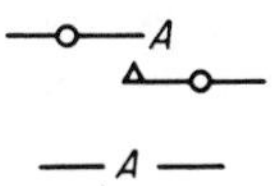 Figure A-15	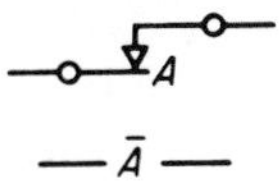 Figure A-16	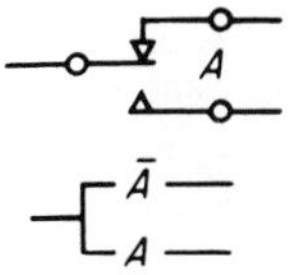 Figure A-17
Springs	2	2	3
Contacts	1	1	2
Positions	1	1	1

Bridge Circuits

Since, in contact networks, the AND function is implemented by series paths and the OR function by parallel paths, any Boolean expression can be *directly* implemented only by a series-parallel network. Frequently, economy can be achieved by the use of *bridge circuits*. A bridge circuit is one in which there is at least one cross-connecting contact between two series paths (Fig. A-18).

A cross-connecting contact in a bridge circuit often conducts current in both directions: for example, the E contact in the A-E-D and C-E-B paths in the circuit above. This is not a necessary requirement, however, as Fig. A-19 shows.

Boolean expression for
series – parallel equivalent:
$AB+CD+ADE+BCE$

Figure A-18

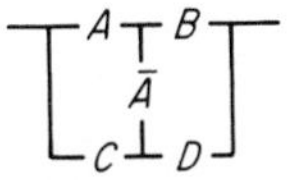

Boolean expression for
series – parallel equivalent:
$AB+CD+BC$

Figure A-19

The circuit in Fig. A-19 is a bridge even though the cross-connecting $\bar{A}$ contact conducts current in only one direction. Current is prevented from flowing in the other direction because of the A and $\bar{A}$ contacts in series.

Nonplanar Networks

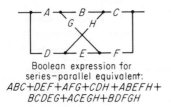

Boolean expression for
series−parallel equivalent:
$ABC+DEF+AFG+CDH+ABEFH+$
$BCDEG+ACEGH+BDFGH$

Figure A-20

Economy is also sometimes achieved by the use of *nonplanar networks* (Fig. A-20). A nonplanar network is one that *cannot* be drawn on a plane without crossing lines. The fact that a circuit is drawn with crossovers does not necessarily make it a nonplanar network since it may be possible to redraw the circuit to eliminate the crossovers. Only when it is impossible to draw the circuit without crossovers is the network nonplanar.

Complementation of Contact Networks

If a contact or two-terminal contact network is closed, its complementary network is open, and vice versa. The complement of a two-terminal series-parallel network can be obtained by

1. Changing all N/O contacts to N/C contacts, and vice versa.
2. Changing all series connections to parallel connections, and vice versa.

For example, the complementary network of Fig. A-21 is Fig. A-22.

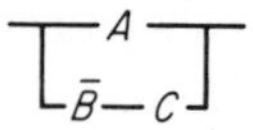

$$A+\bar{B}C$$

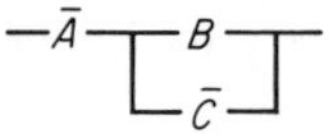

$$\bar{A}(B+\bar{C})$$

Figure A-21   **Figure A-22**

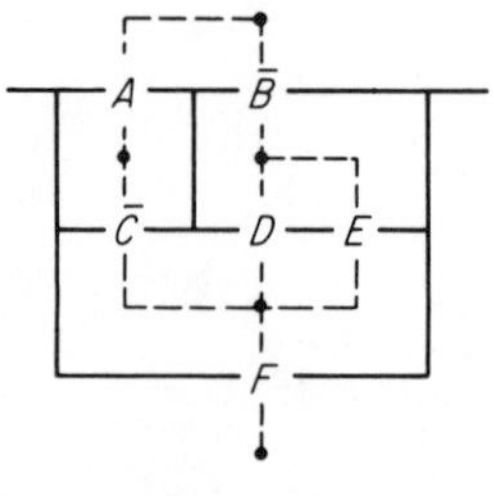

There is a graphical method, however, for obtaining the complement of any planar two-terminal contact network, including bridges. First, a *mesh* in a contact network will be defined as a closed loop that does not contain any smaller loop. Thus, in Fig. A-23, the loop containing contacts A and $\bar{C}$ is a mesh; the loop containing $\bar{B}$, D, and E is a mesh; and the loop containing contacts $\bar{C}$, D, E, and F is a mesh. In addition, the area above the network

Figure A-23

is considered a mesh, as is the area below the network. Thus, the circuit in Fig. A-23 has five meshes. A point, or *node*, is placed in each mesh as shown. The nodes in adjacent meshes are connected with lines passing through contacts common to both meshes. This is done in all possible ways. These connecting lines must always pass through contacts; they may never "cut a wire." The input and output terminals are considered as extending to infinity, so that a connection cannot "circle around" an input or output terminal. The progress at this point is shown in Fig. A-23.

To obtain the complementary network, the new connections are retained and the original connections are deleted, and all contacts are complemented; that is, all *N/O* contacts are changed to *N/C* contacts and vice versa. The top and bottom nodes become the input and output terminals of the complementary network. The resultant complementary network is shown in Fig. A-24.

Figure A-25 is an example of graphical complementation applied to a bridge circuit.

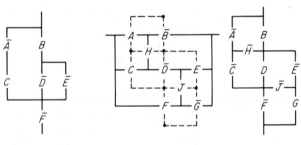

Figure A-24   **Figure A-25**

The number of contacts required for a complementary network is always the same as that required for the original network, although the spring and position count may differ.

A nonplanar network must be converted to a planar equivalent before graphical complementation can be applied.

Relay Trees

A *tree* is a multi-output circuit in which each input combination has a unique output associated with it. A relay tree, or transfer tree, is a particular type of multi-terminal relay contact network having a single input which may be connected to any one of a number of outputs.[2] Only one output is

---

[2]Trees are also sometimes used in reverse; that is, one of a number of inputs is connected to a single output.

connected to the input at any given time, the selection being controlled by the combination of relays operated. Each input-to-output path passes through one contact on each relay, and all outputs are disjunctive; that is, no output can ever be connected to another output through the circuit.

The number of possible relay combinations with n relays is 2^n; therefore, in an n-relay tree there are 2^n possible outputs. A full tree has an output terminal for each of the 2^n possible relay combinations, and the total number of transfers in the tree is $2^n - 1$. A partial tree has less than 2^n output terminals, and the total number of transfers in the tree may vary. A full transfer tree is shown in Fig. A-26; the transfer contact distribution is

$$\begin{array}{cccc} A & B & C & D \\ \hline 1 & 2 & 4 & 8 = 15 \end{array}$$

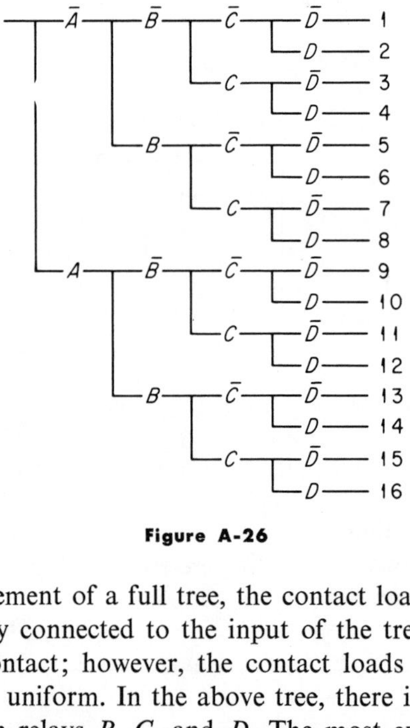

Figure A-26

By the rearrangement of a full tree, the contact load may be somewhat equalized. The relay connected to the input of the tree must always have a single transfer contact; however, the contact loads on the other relays may be made more uniform. In the above tree, there is a total of fourteen transfer contacts on relays B, C, and D. The most even contact division among these three relays is a 4—5—5 distribution.

A circuit with the transfer contact distribution

$$\begin{array}{cccc} A & B & C & D \\ \hline 1 & 4 & 5 & 5 \end{array}$$

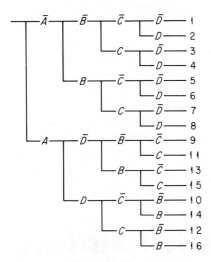

Figure A-27

is shown in Fig. A-27. The outputs are numbered to correspond with those in the previous tree.

Regardless of the contact distribution, the total number of transfers in a full tree is always $2^n - 1$. Rearrangement of a partial tree, however, can lead to a reduction in the total number of transfers required. A discussion on the minimization of partial trees can be found in the literature.

B

Symmetric Functions

Design a circuit that will be closed if and only if exactly three out of a total of eight relays are operated.

An understanding of symmetric functions is useful in the design of switching circuits, particularly relay contact networks (see the example above), where symmetric switching functions lead directly to bridge and nonplanar networks that are much more economical than the best series-parallel circuit otherwise obtainable.

Variables of Symmetry

The function

$$XY\bar{Z} + X\bar{Y}Z + \bar{X}YZ$$

is said to be symmetric in X, Y, and Z since the successive interchanges of any two of the variables X, Y, and Z leaves the function unaltered. The interchanging of X and Z, for example, that is, the replacement of all X's with Z's, all $\bar{X}$'s with $\bar{Z}$'s, all Z's with X's, and all $\bar{Z}$'s with $\bar{X}$'s, results in

$$ZY\bar{X} + Z\bar{Y}X + \bar{Z}YX$$

246

which is identical with the original function. X, Y, and Z in this function are called the *variables of symmetry*. A symmetric function is defined as one in which the interchange of any of the variables of symmetry leaves the function identically the same.

In the preceding function, all variables of symmetry were uncomplemented. Sometimes, in a symmetric function, some of the variables of symmetry may be complemented. For example, the function

$$X\bar{Y}Z + \bar{X}YZ + \bar{X}\bar{Y}\bar{Z}$$

is symmetric in X, Y, and $\bar{Z}$, that is, X, Y, and $\bar{Z}$ are the variables of symmetry. Again, the interchanging of any two of the variables of symmetry will result in the identical function. For instance, interchanging X and $\bar{Z}$ (replacing all X's with $\bar{Z}$'s, all $\bar{X}$'s with Z's, all Z's with $\bar{X}$'s and all $\bar{Z}$'s with X's) results in

$$\bar{Z}\bar{Y}\bar{X} + ZY\bar{X} + Z\bar{Y}X$$

which is identical to the original expression.

m-out-of-n Functions

Symmetric functions in which all of the variables of symmetry are uncomplemented are commonly called *m-out-of-n* functions. Algebraically, these functions equal 1 if exactly m out of the n variables equal 1.

For example, the function

$$\bar{A}BC + A\bar{B}C + AB\bar{C}$$

can be described as a "symmetric 2-out-of-3 function of the variables, A, B, and C," and can be written in "symmetric notation" as

$$S_2^3 ABC$$

A, B, and C are the variables of symmetry, and the expression will equal 1 when exactly two of the three variables equal 1 and under no other conditions.

Symmetric functions may be defined by multiple m's. For example, the function

$$XY + XZ + YZ$$

equals 1 only if two or three of the variables equal 1. This function can be written in symmetric notation as

$$S_{2,3}^3 XYZ$$

and can be described as a symmetric 2- or 3-out-of-3 function of the variables X, Y, and Z.

Boolean Operations with Symmetric Notations

Boolean operations can be performed with symmetric notations; that is, expressions with the same variables of symmetry can be ANDed and ORed, and these expressions or the variables of symmetry or both can be complemented. First, the ANDing of symmetric functions will be examined.

EXAMPLE

$$(S_{1,2,4}^5 ABCDE)(S_{2,3,4}^5 ABCDE) = S_{2,4}^5 ABCDE$$

$S_{1,2,4}^5 ABCDE$ equals 1 if one, two, or four of the variables of symmetry equal 1. $S_{2,3,4}^5 ABCDE$ equals 1 if two, three, or four of the variables of symmetry equal 1. For the product to equal 1, both terms must equal 1, and this can occur only if either two or four of the variables of symmetry equal 1.

Thus, ANDing two symmetric functions containing the same variables of symmetry is accomplished by retaining those subscripts common to both terms. If there are no subscripts in common, the product, of course, equals 0.

Next, the ORing of symmetric functions will be examined.

EXAMPLE

$$S_{1,2,4}^5 ABCDE + S_{2,3,4}^5 ABCDE = S_{1,2,3,4}^5 ABCDE$$

The sum of the two terms will equal 1 if either term equals 1, that is, if one, two, three, or four of the variables of symmetry equal 1. Thus, in ORing two symmetric functions, all of the subscripts in both terms appear in the final expression. If every possible subscript, from 0 through n, occurs, the sum equals 1.

Now, the complementation of a symmetric function will be discussed.

EXAMPLE

$$\overline{S_{1,2,3,4}^5 ABCDE} = S_{0,5}^5 ABCDE$$

The function $\overline{S_{1,2,3,4}^5 ABCDE}$ is *not* equal to 1 (is equal to 0) if one, two, three, or four of the variables of symmetry equal 1. It is logically equivalent to say that this function equals 1 for any condition other than one, two, three, or four of the variables of symmetry equalling 1. The only other possible conditions are none or five of the variables of symmetry equalling 1.

Thus, complementing a symmetric function is accomplished by supplying all subscripts, from 0 through n, that are missing from the original expression. In the example, the missing subscripts are 0 and 5.

Finally, the operation of *complementing the variables of symmetry* will be examined.

EXAMPLE

$$S_{2,4}^5 ABCDE = S_{1,3}^5 \bar{A}\bar{B}\bar{C}\bar{D}\bar{E}$$

The expression $S_{2,4}^5 ABCDE$ equals 1 if two or four of the variables of symmetry equals 1. Saying that two of the variables of symmetry equals 1 is the same as saying that n minus two (or, in this example, three) of the variables of symmetry equal 0. Saying that four of the variables of symmetry equal 1 is the same as saying that one of the variables of symmetry equal 0.

Therefore, another way of saying that a symmetric function equals 1 if two or four of the five variables of symmetry equal 1 is to say that the function equals 1 if one or three of the five variables of symmetry equal 0. Still another way of saying the same thing is that the function equals 1 if one or three of the *complemented variables of symmetry* ($\bar{A}$, $\bar{B}$, $\bar{C}$, $\bar{D}$, and $\bar{E}$) equal 1.

Thus, another way of writing a symmetric function is to complement all of the variables of symmetry and obtain a new set of subscripts by subtracting each of the original subscripts from the total number of variables.

For practice, the equivalence of the following four symmetric expressions should be verified:

$$S_{1,3,4,5}^5 ABCDE \qquad \overline{S_{0,2}^5 ABCDE}$$

$$S_{0,1,2,4}^5 \bar{A}\bar{B}\bar{C}\bar{D}\bar{E} \qquad \overline{S_{3,5}^5 \bar{A}\bar{B}\bar{C}\bar{D}\bar{E}}$$

Symmetric Relay Contact Networks

Suppose that a relay network is desired that is closed only when exactly three out of a total eight relays are operated. The Boolean expression for this circuit might start out like

$$ABC\bar{D}\bar{E}\bar{F}\bar{G}\bar{H} + AB\bar{C}D\bar{E}\bar{F}\bar{G}\bar{H} + A\bar{B}CD\bar{E}\bar{F}\bar{G}\bar{H} + \dots \text{ etc.}$$

for 56 ($_8C_3$) terms. Examination of this expression will show that no simplification is possible (other than factoring). However, a 3-out-of-8 circuit is a symmetric circuit, and circuits of this type can be designed in a matter of seconds, even though they may be complex bridge networks or even nonplanar networks.

First, the general structure of symmetric networks will be examined.

Symmetric Trees

A symmetric tree is a multi-output relay circuit with one input and $n + 1$ outputs, where n is the total number of relays in the circuit. The outputs are numbered from 0 through n, and with m out of n relays operated, the m output is connected to the input.

As an example, a three-relay symmetric tree is shown in Fig. B-1, both in conventional symbolic form, and also in a diagrammatic form that is convenient to use for symmetric circuits.

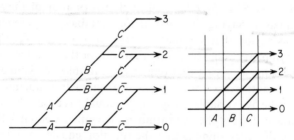

Figure B-1

Referring first to the circuit diagram on the left, it can be seen that with zero relays operated, the input is connected to the 0 output; with one relay operated, the input is connected to the 1 output; etc. It should be obvious that a circuit structure of this type can be extended to include any number of relays. The diagram on the right represents the identical circuit. To construct such a diagram, $n + 1$ vertical guide lines and $n + 1$ horizontal guide lines are drawn, as shown. Then n relay designations are written at the bottom, in the spaces between the vertical guide lines. At the right, the horizontal guide lines are labeled, from bottom to top, with the output designations 0 to n. Horizontal and diagonal lines are then drawn as shown; all horizontal lines between two vertical guide lines represent normally closed contacts on the relay indicated below, and all diagonal lines between the two vertical guide lines represent normally open contacts on that relay.

For the usual symmetric circuit requirement, only a portion of a symmetric tree is required. Suppose, for instance, that a circuit is desired that is closed only when exactly three out of eight relays, A through H, are operated. This circuit requirement can be written in symmetric notation as

$$S_3^8 ABCDEFGH$$

The circuit can immediately be drawn, as shown in Fig. B-2. Nine vertical guide lines are drawn—one more than the total number of relays involved—

leaving a vertical space for each of the
eight relays. Four horizontal guide lines
are drawn—one more than the number
of relays that must be operated for an
output—the topmost guide line representing the 3 (relays operated) output.
Only that portion of the symmetric tree
leading to the 3 output is then drawn.

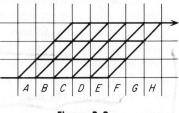

Figure B-2

The order of the relays is arbitrary;
no matter what the order, the circuit diagram remains the same. Also, the
contact distribution cannot be equalized; a relay closer to either end of
the diagram usually requires fewer contacts than one nearer the middle
of the diagram.

Identification of Transfer Contacts

Figure B-3 shows a method for identifying transfer contacts on symmetric
circuit diagrams. A small arc is drawn between a normally open and normally
closed contact, signifying a transfer contact. Six transfer contacts are required
for the symmetric tree in Fig. B-3.

The 3-out-of-8 circuit is redrawn in Fig. B-4, with the transfer contacts
identified. Note that on relays *A* and *H* there is one transfer contact each;

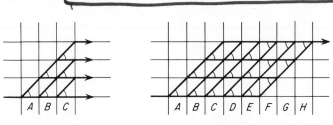

Figure B-3   **Figure B-4**

on relays *B* and *G*, two transfers each; on relays *C* and *F*, three transfers each;
and on relays *D* and *E*, three transfers plus one normally closed contact each.

Information on techniques for designing symmetric circuits with multiple
m's can be found in the literature.

The discussion of symmetric functions has been centered on the *m-out-of-n*
type, in which all of the variables of symmetry are uncomplemented (or all
complemented). The recognition of symmetric functions in which *some* of the
variables of symmetry are complemented is a problem in itself, and will not
be discussed here. Information on the detection and identification of symmetric functions in which any number of the variables of symmetry may be
complemented can be found in the literature.

Once a symmetric function has been detected and identified, the corresponding relay contact network can be designed using the symmetric tree approach modified as follows. If a variable of symmetry is uncomplemented, the corresponding diagonal lines represent normally open contacts and the horizontal lines represent normally closed contacts; if the variable of symmetry is complemented, the meaning of the lines is reversed, the horizontal lines representing normally open contacts and the diagonal lines normally closed contacts.

As an example, the relay contact network for the function $S_3^5 A\bar{B}CD\bar{E}$ is shown in Fig. B-5.

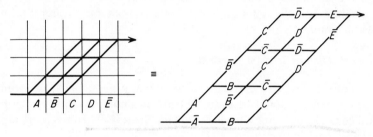

Figure B-5

C

Sequential Relay Circuits

Relay sequential circuits function in level operation. The relay is a memory element, and, in fact, is a Y element (refer to Chapter 12); Y is associated with the relay coil, and y with the relay contact. A signal on the input (coil), Y, at time t, appears on the output (contact), y, at time $t + \Delta$. The delay, Δ, is inherent in the transition time of the relay.

The state of a relay is a function of the operation or inoperation of the contacts; the excitation is a function of the signal or no signal on the coil. Refer to the coil and normally open contact in Fig. C-1:

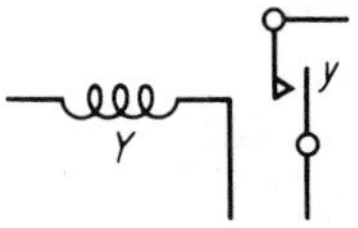

$y = 0$ Contact unoperated

$y = 1$ Contact operated

$Y = 0$ No signal on coil

$Y = 1$ Signal on coil

Figure C-1

When $Y = y$, the next state of the relay will be the same as the present state, and since the relay will not change state it is stable. When $Y \neq y$, the next state will not be the same as the present state, and since the relay will change state it is not stable.

Assume, in Fig. C-1, that there is no signal on the coil and that the contact is unoperated. This is a stable condition: $Y = y = 0$.

Now assume a signal on the coil. For a brief period of time, Δ, the contact is still unoperated. During this time, $Y = 1$ and $y = 0$, and the relay is unstable.

After the transition time of the relay, the contact will operate and the relay will again become stable: $Y = y = 1$.

Next assume that the signal is removed from the coil. For a time, Δ, the contact is still operated. During this time, $Y = 0$ and $y = 1$, and the relay is unstable.

After the transition time of the relay, the contact will return to its unoperated condition, and the relay will again be stable: $Y = y = 0$.

A schematic diagram of a relay sequential circuit is shown in Fig. C-2. The primary relays are under direct control of the inputs, and are used to make a multiplicity of contacts available for switching. The secondary relays are the memory elements.

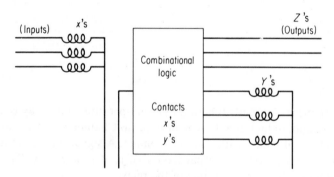

Schematic diagram of relay sequential circuit

Figure C-2

The entire process for synthesizing relay sequential circuits is the same as that for Y elements. Hazards peculiar to relay circuits, and means for their elimination, warrant special mention, however.

Hazards

Consider the relay circuit in Fig. C-3, in which the transfer contacts on relay A are of the break-before-make type.

Figure C-3

Assume a condition of relays B and C operated, and relay A changing state. The circuit is closed before and after the change, but for a brief interval of time during the transition of relay A, both the A and $\bar{A}$ contacts are open, and the circuit is therefore open.

The hazard can be eliminated without adding redundancy by making use of the relationship

$$AB + \bar{A}C = (A + C)(\bar{A} + B)$$

and implementing the circuit as in Fig. C-4. With this implementation, when both the A and $\bar{A}$ contacts are open during the transition of relay A, the circuit remains closed, a path being established through the closed B and C contacts.

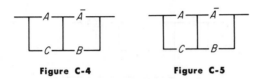

Figure C-4   **Figure C-5**

Now consider the relay implementation in Fig. C-5 in which the transfer contacts on relay A are of the make-before-break or continuity transfer type.

Assume a condition of relays B and C unoperated, and relay A changing state. The circuit is open before and after the change, but for a brief interval of time during the transition of relay A, both the A and $\bar{A}$ contacts are closed, and the circuit is therefore closed.

This hazard can be eliminated without adding redundancy by implementing the circuit as in Fig. C-6. With this implementation, when both the A and $\bar{A}$ contacts are closed during the transition of relay A, the circuit remains open, the open B and C contacts preventing any path from being established.

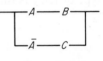

Figure C-6

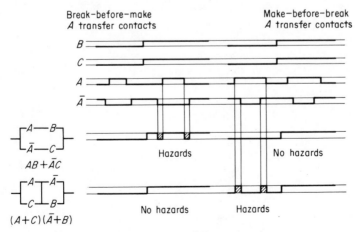

Figure C-7

The two types of hazards just discussed are illustrated in timing charts in Fig. C-7.

In addition to the static hazards discussed, dynamic and essential hazards can also occur with relay circuits.

An implementation of a relay sequential circuit from the expressions

$$Y = x_2 y + x_1 \bar{x}_2$$
$$Z = x_2 y$$

is shown in Fig. C-8.

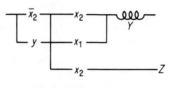

Figure C-8

Another type of relay, the *latch relay*, has two coils and a mechanical latching mechanism. A signal on the "latch" coil causes the relay to mechanically latch in the operated positions. The relay then remains in this state until a signal on the "unlatch" coil releases the mechanism, returning the relay to its normal state. The synthesis of relay sequential circuits with this type of relay is the same as that for *S-C* flip-flops with level operation.

Related Literature[1]

Chapter 1

G. BOOLE, *The Mathematical Analysis of Logic*, Cambridge, 1847.

G. BOOLE, *An Investigation of the Laws of Thought*, London, 1854.

C. E. SHANNON, "A Symbolic Analysis of Relay and Switching Circuits," *Trans. AIEE*, Vol. 57, pp. 713–723, 1938.

W. H. KAUTZ, "A Survey and Assessment of Progress in Switching Theory and Logical Design in the Soviet Union," *IEEETEC*, Vol. EC-15, No. 2, pp. 164–204, April, 1966.

Chapter 2

W. H. BURKHARDT, "Theorem Minimization," *Proceedings of the Assoc. for Computing Machinery*, pp. 259–263, May 2–3, 1952.

V. KUDIELKA and P. OLIVA, "Complete Sets of Functions of Two and Three Binary Variables," *IEEETEC*, Vol. EC-15, No. 6, pp. 930–931, Dec., 1966.

F. M. BROWN, "Reduced Solutions of Boolean Equations," *IEEETC*, Vol. C-19, No. 10, pp. 976–981, Oct., 1970.

---

[1] Chronological by chapter.

R. V. SETLUR, "A Method to Determine the Expressive Power of a Set of Connectives," *IEEETC*, Vol. C-19, No. 12, pp. 1223–1225, Dec., 1970.

F. P. PREPARATA, "On the Design of Universal Boolean Functions," *IEEETC*, Vol. C-20, No. 4, pp. 418–423, April, 1971.

S. Y. H. SU and A. A. SARRIS, "The Relationship Between Multivalued Switching Algebra and Boolean Algebra Under Different Definitions of Complement," *IEEETC*, Vol. C-21, No. 5, pp. 479–485, May, 1972.

Chapter 3

S. H. WASHBURN, "An Application of Boolean Algebra to the Design of Electronic Switching Circuits," *AIEE, Part I, Communication and Electronics*, Vol. 72, pp. 380–388, Sept., 1953.

E. C. NELSON, "An Algebraic Theory for Use in Digital Computer Design," *IRETEC*, Vol. EC-3, No. 3, pp. 12–21, Sept., 1954.

B. J. YOKELSON and W. ULRICH, "Engineering Multi-Stage Diode Logic Circuits," *AIEE, Communication and Electronics*, No. 20, pp. 466–475, Sept., 1955.

T. J. BEATSON, "Minimization of Components in Electronic Switching Circuits," *AIEE, Part I, Communication and Electronics*, Vol. 77, pp. 283–291, July, 1958.

"Military Standard Graphic Symbols for Logic Diagrams," *MIL-STD-806B*, Feb. 26, 1962.

"American Standard Graphic Symbols for Logic Diagrams," *American Standards Association, ASA Y32.14-1962*, Sept. 26, 1962. (Published by the *AIEE*.)

Chapter 4

W. V. QUINE, "The Problem of Simplifying Truth Functions," *American Mathematical Monthly*, Vol. 59, pp. 521–531, Oct., 1952.

E. W. SAMSON and B. E. MILLS, "Circuit Minimization: Algebra and Algorithm for new Boolean Canonical Expressions," *Air Force Cambridge Research Center Technical Report 54–21*, April, 1954.

D. E. MULLER, "Application of Boolean Algebra to Switching Circuit Design and to Error Detection," *IRETEC*, Vol. EC-3, No. 3, pp. 6–12, Sept., 1954.

R. MUELLER, "On the Synthesis of a Minimal Representation of a Logic Function," *Air Force Cambridge Research Center Technical Report 55-104*, April 1955.

R. J. NELSON, "Simplest Normal Truth Functions," *Journal of Symbolic Logic*, Vol. 20, No. 2, pp. 105–108, June, 1955.

E. W. SAMSON and R. MUELLER, "Circuit Minimization: Minimal and Irredundant Boolean Sums by Alternative Set Method," *Air Force Cambridge Research Center Technical Report 55-109*, June, 1955.

R. J. NELSON, "Weak Simplest Normal Truth Functions," *Journal of Symbolic Logic*, Vol. 20, No. 3, pp. 232–234, Sept., 1955.

W. V. QUINE, "A Way to Simplify Truth Functions," *American Mathematical Monthly*, Vol. 62, pp. 627–631, Nov., 1955.

D. E. MULLER, "Complexity in Electronic Switching Circuits," *IRETEC*, Vol. EC-5, No. 1, pp. 15–19, March, 1956.

S. R. PETRICK, "A Direct Determination of the Irredundant Forms of a Boolean Function from the Set of Prime Implicants," *Air Force Cambridge Research Center Technical Report 56-110*, April, 1956.

R. H. URBANO and R. K. MUELLER, "A Topological Method for the Determination of the Minimal Forms of a Boolean Function," *IRETEC*, Vol. EC-5, No. 3, pp. 126–132, Sept., 1956.

E. J. McCLUSKEY, JR., "Minimization of Boolean Functions," *BSTJ*, Vol. 35, No. 6, pp. 1417–1444, Nov., 1956.

M. J. GHAZALA (also GAZALÉ), "Irredundant Disjunctive and Conjunctive Forms of a Boolean Function," *IBM Journal of Research and Development*, Vol. 1, No. 2, pp. 171–176, April, 1957.

T. SINGER, "Some Uses of Truth Tables," *Proceedings of an International Symposium on the Theory of Switching, Part I*, Harvard University, Cambridge, Mass. pp. 125–133, April, 1957.

B. HARRIS, "An Algorithm for Determining Minimal Representations of a Logic Function," *IRETEC*, Vol. EC-6, No. 2, pp. 103–108, June, 1957.

R. McNAUGHTON and B. MITCHELL, "The Minimality of Rectifier Nets with Multiple Outputs Incompletely Specified," *Journal of the Franklin Institute*, Vol. 264, No. 6, pp. 457–480, Dec., 1957.

J. N. WARFIELD, "A Note on the Reduction of Switching Functions," *IRETEC*, Vol. EC-7, No. 2, pp. 180–181, June, 1958.

T. C. BARTEE, "The Automatic Design of Logical Networks," *Proc. of the Western Joint Computer Conference*, pp. 103–107, March 3–5, 1959. (Published by the *IRE*.)

B. DUNHAM and R. FRIDSHAL, "The Problem of Simplifying Logical Expressions," *Journal of Symbolic Logic*, Vol. 24, No. 1, pp. 17–19, March, 1959.

J. P. ROTH, "Algebraic Topological Methods in Synthesis," *Proceedings of an International Symposium on the Theory of Switching, Part I*, Harvard University, Cambridge, Mass., pp. 57–73, April, 1959.

R. B. POLANSKY, "Further Notes on Simplifying Multiple-output Switching Circuits," *Electronics Systems Laboratory Mem. 7849-M-330*, M.I.T., Cambridge, Mass., pp. 1–6, Oct. 26, 1959.

J. P. ROTH and E. G. WAGNER, "Algebraic Topological Methods for the Synthesis of Switching Systems, Part III: Minimization of Nonsingular Boolean Trees," *IBM Journal of Research and Development*, Vol. 3, No. 4, pp. 326–344, Oct., 1959.

W. V. QUINE, "On Cores and Prime Implicants of Truth Functions," *American Mathematical Monthly*, Vol. 66. pp. 755–760, Nov., 1959.

T. H. MOTT, JR., "Determination of the Irredundant Normal Forms of a Truth Function by Iterated Consensus of the Prime Implicants," *IRETEC*, Vol. EC-9, No. 2, pp. 245–252, June, 1960.

J. P. Rотн, "Minimization Over Boolean Trees," *IBM Journal of Research and Development*, Vol. 4, No. 5, pp. 543–558, Nov., 1960.

G. C. Vandling, "The Simplification of Multiple-Output Networks Composed of Unilateral Devices," *IRETEC*, Vol. EC-9, No. 4, pp. 477, 486, Dec., 1960.

T. C. Bartee, "Computer Design of Multiple-Output Logical Networks," *IRETEC*, Vol. EC-10, No. 1, pp. 21–30, March, 1961.

R. B. Polansky, "Minimization of Multiple-Output Switching Circuits," *AIEE Transactions, Part I, Communication and Electronics*, Vol. 80, pp. 67–73, March, 1961.

J. T. Chu, "A Generalization of a Theorem of Quine for Simplifying Truth Functions," *IRETEC*, Vol. EC-10, No. 2, pp. 165–168, June, 1961.

E. J. McCluskey, Jr. "Minimal Sums for Boolean Functions Having Many Unspecified Fundamental Products," *Proceedings of the Second Annual Symposium on Switching Circuit Theory and Logical Design*, pp. 10–17, Sept., 1961. (Published by the *AIEE*.)

S. B. Akers, Jr., "A Truth Table Method for the Synthesis of Combinational Logic," *IRETEC*, Vol. EC-10, No. 4, pp. 604–615, Dec., 1961.

F. B. Hall, "Boolean Prime Implicants by the Binary Sieve Method," *AIEE Transactions, Part I, Communication and Electronics*, Vol. 80, pp. 709–713, Jan., 1962.

R. Hockney, "An Intersection Algorithm Giving All Irredundant Forms from a Prime Implicant List," *IEEETEC*, Vol. EC-11, No. 2, pp. 289–290, April, 1962.

T. Rado, "Comments on the Presence Function of Gazalé," *IBM Journal of Research and Development*, Vol. 6, No. 2, pp. 268–269, April, 1962.

J. P. Roth and R. M. Karp, "Minimization Over Boolean Graphs," *IBM Journal of Research and Development*, Vol. 6, No. 2, pp. 227–238, April, 1962.

Z. Kohavi, "Minimizing of Incompletely Specified Sequential Switching Circuits," *Office of Technical Services Government Research Report AD 286,174*, May 10, 1962.

A. H. Scheinman, "A Method for Simplifying Boolean Functions," *BSTJ*, Vol. 41, No. 4, pp. 1337–1346, July, 1962.

I. B. Pyne and E. J. McCluskey, Jr. "The Reduction of Redundancy in Solving Prime Implicant Tables," *IRETEC*, Vol. EC-11, No. 4, pp. 473–482, Aug., 1962.

A. K. Choudhury and M. S. Basu, "A Mechanized Chart for Simplification of Switching Functions," *IRETEC*, Vol. EC-11, No. 5, pp. 713–714, Oct., 1962.

F. Mileto and G. Putzolu, "Average Values of Quantities Appearing in Boolean Function Minimization," *IEEETEC*, Vol. EC-13, No. 2, pp. 87–92, April, 1964.

H. Mott and C. C. Carroll, "Numerical Procedures for Boolean Function Minimization," *IEEETEC*, Vol. EC-13, No. 4, p. 470, Aug., 1964.

R. S. Gaines, "Implication Techniques for Boolean Functions," *Proceedings of the Fifth Annual Symposium on Switching Circuit Theory and Logical Design*, S-164, pp. 174–182, Oct., 1964. (Published by the *IEEE*.)

J. F. GIMPEL, "A Reduction Technique for Prime Implicant Tables," *Proceedings of the Fifth Annual Symposium on Switching Circuit Theory and Logical Design,* S-164, pp. 183–191, Oct., 1964. (Published by the *IEEE.*)

D. M. Y. CHANG and T. H. MOTT, JR. "Computing Irredundant Normal Forms from Abbreviated Presence Functions," *IEEETEC,* Vol. EC-14, No. 3, pp. 335–342, June, 1965.

J. F. GIMPEL, "A Method of Producing a Boolean Function Having an Arbitrarily Prescribed Prime Implicant Table," *IEEETEC,* Vol. EC-14, No. 3, pp. 485–488, June, 1965.

A. J. NICHOLS and A. J. BERNSTEIN, "State Assignments in Combinational Networks," *IEEETEC,* Vol. EC-14, No. 3, pp. 343–349, June, 1965.

J. F. GIMPEL, "A Reduction Technique for Prime Implicant Tables," *IEEETEC,* Vol. EC-14, No. 4, pp. 535–541, Aug., 1965.

F. MILETO and G. PUTZOLU, "Average Values of Quantities Appearing in Multiple Output Boolean Minimization," *IEEETEC,* Vol. EC-14, No. 4, pp. 542–552, Aug., 1965.

S. R. DAS and A. K. CHOUDHURY, "Maxterm Type Expressions of Switching Functions and Their Prime Implications, "*IEEETEC,* Vol. EC-14, No. 6, pp. 920–923, Dec., 1965.

A. R. MEO, "On the Determination of the *ps* Maximal Implicants of a Switching Function," *IEEETEC,* Vol. EC-14, No. 6, pp. 830–840, Dec., 1965.

F. LUCCIO, "A Method for the Selection of Prime Implicants," *IEEETEC,* Vol. EC-15, No. 2, pp. 205–212, April, 1966.

A. K. CHOUDHURY and S. R. DAS, "Computing Irredundant Normal Forms from Abbreviated Presence Functions," *IEEETEC,* Vol. EC-15, No. 3, p. 387, June, 1966.

R. D. MERRILL, JR., "A Tabular Minimization Procedure for Ternary Switching Functions," *IEEETEC,* Vol. EC-15, No. 4, pp. 578–585, Aug., 1966.

B. B. GORDON et al., "Simplification of the Covering Problem for Multiple Output Logical Networks," *IEEETEC,* Vol. EC-15, No. 6, pp. 891–897, Dec., 1966.

S. U. ROBINSON III and R. W. HOUSE, "Gimpel's Reduction Technique Extended to the Covering Problem with Costs," *IEEETEC,* Vol. EC-16, No. 4, pp. 509–514, Aug., 1967.

P. TISON, "Generalization of Consensus Theory and Application to the Minimization of Boolean Functions," *IEEETEC,* Vol. EC-16, No. 4, pp. 446–456, Aug., 1967.

E. MORREALE, "Partitioned List Algorithms for Prime Implicant Determination from Canonical Forms," *IEEETEC,* Vol. EC-16, No. 5, pp. 611–620, Oct., 1967.

N. N. NECULA, "A Numerical Procedure for the Determination of the Prime Implicants of a Boolean Function, "*IEEETEC,* Vol. EC-16, No. 5, pp. 687–689, Oct., 1967.

C. M. ALLEN and D. D. GIVONE, "A Minimization Technique for Multiple-Valued Logic Systems," *IEEETC,* Vol. C-17, No. 2, pp. 182–184, Feb., 1968.

P. R. Schneider and D. L. Dietmeyer, "An Algorithm for Synthesis of Multiple-Output Combinational Logic," *IEEETC*, Vol. C-17, No. 2, pp. 117–128, Feb., 1968.

N. N. Necula, "An Algorithm for the Automatic Approximate Minimization of Boolean Functions," *IEEETC*, Vol. C-17, No. 8, pp. 770–782, Aug., 1968.

E. S. Davidson and G. Metze, "Comments on 'An Algorithm for Synthesis of Multiple-Output Combinational Logic,'" *IEEETC*, Vol. C-17, No. 11, pp. 1091–1092, Nov., 1968.

Y. H. Su and D. L. Dietmeyer, "Computer Reduction of Two-Level Multiple-Output Switching Circuits," *IEEETC*, Vol. C-18, No. 1, pp. 58–63, Jan., 1969.

J. M. Mage, "Application of Iterative Consensus to Multiple-Output Functions," *IEEETC*, Vol. C-19, No. 4, p. 359, April, 1970.

E. Morreale, "Recursive Operators for Prime Implicant and Irredundant Normal Form Determination," *IEEETC*, Vol. C-19, No. 6, pp. 504–509, June, 1970.

M. P. Marcus and W. H. Niehoff, "Iterated Consensus Method for Multiple-output Functions," *IBM Journal of Research and Development*, Vol. 14, No. 6, pp. 677–679, Nov., 1970.

P. L. Tison, "An Algebra for Logic Systems—Switching Circuits Application," *IEEETC*, Vol. C-20, No. 3, pp. 339–351, March, 1971.

J. G. Bredeson, "Generation of Prime Implicants by Direct Multiplication," Vol. C-20, No. 4, pp. 475–476, April, 1971.

D. D. Givone, M. E. Liebler, and R. P. Roesser, "A Method of Solution for Multiple-Valued Logic," *IEEETC*, Vol. C-20, No. 4, pp. 464–467, April, 1971.

N. N. Biswas, "Minimization of Boolean Functions," *IEEETC*, Vol. C-20, No. 8, pp. 925–929, Aug., 1971.

S. R. Das and N. S. Khabra, "Clause-Column Table Approach for Generating All the Prime Implicants of Switching Functions," *IEEETC*, Vol. C-21, No. 11, pp. 1239–1246, Nov., 1972.

V. Bubenik, "Weighting Method for the Determination of the Irredundant Set of Prime Implicants," *IEEETC*, Vol. C-21, No. 12, pp. 1449–1451, Dec., 1972.

Chapter 5

E. W. Veitch, "A Chart Method for Simplifying Truth Functions," *Proceedings of the Assoc. for Computing Machinery*, pp. 127–133, May 2–3, 1952.

M. Karnaugh, "The Map Method for Synthesis of Combinational Logic Circuits," *Trans. AIEE, Part I, Communication and Electronics*, Vol. 72, pp. 593–599, Nov., 1953.

M. E. Arthur, "Geometric Mapping of Switching Functions," *IRETEC*, Vol. EC-10, No. 4, pp. 631–637, Dec., 1961.

T. M. Booth, "The Vertex-Frame Method for Obtaining Minimal Proposition-Letter Formulas," *IRETEC*, Vol. EC-11, No. 2, pp. 144–154, April, 1962.

G. E. Marthugh and R. E. Anderson, "The H Diagram: A Graphical Approach to Logic Design," *IEEETC*, Vol. C-20, No. 10, pp. 1192–1196, Oct., 1971.

Chapter 6

R. C. Brigham, "Some Properties of Binary Counters with Feedback," *IRETEC*, Vol. EC-10, No. 4, pp. 699–701, Dec., 1961.

M. P. Marcus, "Cascaded Binary Counters with Feedback," *IEEETEC*, Vol. EC-12, No. 4, pp. 361–364, Aug., 1963.

Chapter 7

R. W. Hamming, "Error Detecting and Error Correcting Codes," *BSTJ*, Vol. 29, No. 2, pp. 147–160, April, 1950.

W. W. Peterson, *Error-Correcting Codes*, The M.I.T. Press and John Wiley & Sons, Inc., New York, 1961.

M. P. Marcus, "Minimum Polarized Distance Codes," *IBM Journal of Research and Development*, Vol. 5, No. 3, pp. 241–248, July, 1961.

Chapter 8

A. W. Burks and J. B. Wright, "Theory of Logical Nets," *Proc. IRE*, Vol. 41, No. 10, pp. 1357–1365, Oct., 1953.

D. A. Huffman, "The Synthesis of Sequential Circuits," *Journal of the Franklin Institute*, Vol. 257, No. 3, pp. 161–190, March, 1954; No. 4, pp. 275–303, April, 1954.

G. H. Mealy, "A Method for Synthesizing Sequential Circuits," *BSTJ*, Vol. 34, No. 5, pp. 1045–1079, Sept., 1955.

N. Zierler, "Several Binary-Sequence Generators," *Lincoln Lab. Technical Report 95*, M.I.T., Sept. 12, 1955.

E. F. Moore, "Gedanken-Experiments on Sequential Machines," *Automata Studies*, Princeton University Press, Princeton, N.J., pp. 129–153, 1956.

D. E. Muller and W. S. Bartky, "A Theory of Asynchronous Circuits," *Proceedings of an International Symposium on the Theory of Switching, Part I*, Harvard University, Cambridge, Mass., pp. 204–243, April, 1957.

S. H. Unger, "A Study of Asynchronous Logical Feedback Networks," *Research Lab. of Electronics Technical Report 320*, M.I.T., April 26, 1957.

A. W. Burks and H. Wang, "The Logic of Automata, Parts I and II," *JACM*, Vol. 4, No. 2, pp. 193–218, April, 1957; No. 3, pp. 279–297, July, 1957.

J. M. Simon, "Some Aspects of the Network Analysis of Sequence Transducers," *Journal of the Franklin Institute*, Vol. 265, No. 6, pp. 439–450, June, 1958.

B. Elspas, "The Theory of Autonomous Linear Sequential Networks," *IRETCT*, Vol. CT-6, No. 1, pp. 45–60, March, 1959.

B. FRIEDLAND, "Linear Modular Sequential Circuits," *IRETCT*, Vol. CT-6, No. 1, pp. 61–68, March, 1959.

J. HARTMANIS, "Linear Multivalued Sequential Coding Networks," *IRETCT*, Vol. CT-6, No. 1, pp. 69–74, March, 1959.

M. O. RABIN and D. SCOTT, "Finite Automata and Their Decision Problems," *IBM Journal of Research and Development*, Vol. 3, No. 2, pp. 114–125, April, 1959.

J. A. BRZOZOWSKI, "A Survey of Regular Expressions and Their Applications," *IRETEC*, Vol. EC-11, No. 3, pp. 324–335, June, 1962.

F. S. STĂNCIULESCU, "Sequential Logic and Its Application to the Synthesis of Finite Automata," *IEEETEC*, Vol. EC-14, No. 6, pp. 786–791, Dec., 1965.

A. D. FRIEDMAN, "Feedback in Synchronous Sequential Switching Circuits," *IEEETEC*, Vol. EC-15, No. 3, pp. 354–367, June, 1966.

S. S. YAU and K. C. WANG, "Linearity of Sequential Machines," *IEEETEC*, Vol. EC-15, No. 3, pp. 337–354, June, 1966.

A. D. FRIEDMAN, "Feedback in Asynchronous Sequential Circuits, *IEEETEC*, Vol. EC-15, No. 5, p. 740, Oct., 1966.

J. HARTMANIS, "Minimal Feedback Realizations of Sequential Machines," *IEEETEC*, Vol. EC-15, No. 6, pp. 931–933, Dec., 1966.

S. S. YAU and K. C. WANG, "Correction to 'Linearity of Sequential Machines,'" *IEEETEC*, Vol. EC-15, No. 6, p. 925, Dec., 1966.

Z. KOHAVI and P. LAVALLEE, "Design of Sequential Machines with Fault Detection Capabilities," *IEEETEC*, Vol. EC-16, No. 4, pp. 473–484, Aug., 1967.

D. R. ARMSTRONG, A. D. FRIEDMAN, and P. R. MENON, "Design of Asynchronous Circuits Assuming Unbounded Gate Delays," *IEEETC*, Vol. C-18, No. 12, pp. 1110–1120, Dec., 1969.

T. F. ARNOLD, C. J. TAN, and M. M. NEWBORN, "Iteratively Realized Sequential Circuits," *IEEETC*, Vol. C-19, No. 1, pp. 54–66, Jan., 1970.

W. A. DAVIS, "Sequential Machines Realizable with Delay Elements Only," *IEEETC*, Vol. C-19, No. 4, pp. 353–355, April, 1970.

B. BEIZER, "Towards a New Theory of Sequential Switching Networks," *IEEETC*, Vol. C-19, No. 10, pp. 939–956, Oct., 1970.

P. J. MARINO, "A Linear Decomposition for Sequential Machines," *IEEETC*, Vol. C-19, No. 10, pp. 956–963, Oct., 1970.

F. P. PREPARATA and D. E. MULLER, "On the Delay Required to Realize Boolean Functions," *IEEETC*, Vol. C-20, No. 4, pp. 459–461, April, 1971.

S. H. UNGER, "Asynchronous Sequential Switching Circuits with Unrestricted Input Changes," *IEEETC*, Vol. C-20, No. 12, pp. 1437–1444, Dec., 1971.

Chapter 9

G. OTT and N. H. FEINSTEIN, "Design of Sequential Machines from their Regular Expressions," *JACM*, Vol. 8, pp. 585–600, Oct., 1961.

J. A. BRZOZOWSKI and E. J. McCLUSKEY, JR., "Signal Flow Graph Techniques for Sequential Circuit State Diagrams," *IEEETEC*, Vol. EC-12, No. 2, pp. 67–76, April, 1963.

A. D. FRIEDMAN and P. R. MENON, "Synthesis of Asynchronous Sequential Circuits with Multiple-Input Changes," *IEEETC*, Vol. C-17, No. 6, pp. 559–566, June, 1968.

R. M. KLINE and D. F. WANN, "Threshold Logic Design of Pulse-Type Sequential Networks," *IEEETC*, Vol. C-18, No. 5, pp. 459–465, May, 1969.

G. K. MAKI, J. H. TRACEY, and R. J. SMITH, "Generation of Design Equations in Asynchronous Sequential Circuits," *IEEETC*, Vol. C-18, No. 5, pp. 467–472, May, 1969.

M. P. MARCUS, "Constructing the Primitive Flow Table," *Electronic Design*, Vol. 25, pp. 62–64, Dec. 9, 1971.

Chapter 10

D. D. AUFENKAMP and F. E. HOHN, "Analysis of Sequential Machines," *IRETEC*, Vol. EC-6, No. 4, pp. 276–285, Dec., 1957.

D. D. AUFENKAMP, "Analysis of Sequential Machines II," *IRETEC*, Vol. EC-7, No. 4, pp. 299–306, Dec., 1958.

W. J. CADDEN, "Equivalent Sequential Circuits," *IRETCT*, Vol. CT-6, No. 1, pp. 30–34, March, 1959.

S. GINSBURG, "A Synthesis Technique for Minimal State Sequential Machines," *IRETEC*, Vol. EC-8, No. 1, pp. 13–24, March, 1959.

S. SESHU, R. E. MILLER, and G. METZE, "Transition Matrices of Sequential Machines," *IRETCT*, Vol. CT-6, No. 1, pp. 5–12, March, 1959.

J. M. SIMON, "A Note on Memory Aspects of Sequence Transducers," *IRETCT*, Vol. CT-6, No. 1, pp. 26–29, March, 1959.

D. B. NETHERWOOD, "Minimal Sequential Machines," *IRETEC*, Vol. EC-8, No. 3, pp. 339–345, Sept., 1959.

M. C. PAULL and S. H. UNGER, "Minimizing the Number of States in Sequential Switching Functions," *IRETEC*, Vol. EC-8, No. 3, pp. 356–367, Sept., 1959.

S. GINSBURG, "A Technique for the Reduction of a Given Machine to a Minimal-State Machine," *IRETEC*, Vol. EC-8, No. 3, pp. 346–355, Sept., 1959.

S. GINSBURG, "Synthesis of Minimal-State Machines," *IRETEC*, Vol. EC-8, No. 4, pp. 441–449, Dec., 1959.

J. HARTMANIS, "Symbolic Analysis of a Decomposition of Information Processing Machines," *Information and Control*, Vol. 3, No. 2, pp. 154–178, June, 1960.

A. GILL, "A Note on Moore's Distinguishability Theorem," *IRETEC*, Vol. EC-10, No. 2, pp. 290–291, June, 1961.

R. NARASIMHAN, "Minimizing Incompletely Specified Sequential Switching Functions," *IRETEC*, Vol. EC-10, No. 3, pp. 531–532, Sept., 1961.

E. J. McCluskey, Jr., "Minimum-State Sequential Circuits for a Restricted Class of Incompletely Specified Flow Tables," *BSTJ*, Vol. 41, No. 6, pp. 1759–1768, Nov., 1962.

J. Hartmanis, "Further Results on the Structure of Sequential Machines," *JACM*, Vol. 10, No. 1, pp. 78–88, Jan., 1963.

J. Hartmanis, "The Equivalence of Sequential Machine Models, "*IEEETEC*, Vol. EC-12, No. 1, pp. 18–19, Feb., 1963.

M. P. Marcus, "Derivation of Maximal Compatibles Using Boolean Algebra," *IBM Journal of Research and Development*, Vol. 8, No. 5, pp. 537–538, Nov., 1964.

I. S. Reed, "Some Remarks on State Reduction of Asynchronous Circuits by the Paull-Unger Method," *IEEETEC*, Vol. EC-14, No. 2, pp. 262–265, April, 1965.

A. Grasselli and F. Luccio, "A Method for Minimizing the Number of Internal States in Incompletely Specified Sequential Networks," *IEEETEC*, Vol. EC-14, No. 3, pp. 350–359, June, 1965.

S. H. Unger, "Flow Table Simplification—Some Useful Aids," *IEEETEC*, Vol. EC-14, No. 3, pp. 472–475, June, 1965.

A. Grasselli, "Minimal Closed Partitions for Incompletely Specified Flow Tables," *IEEETEC*, Vol. EC-15, No. 2, pp. 245–249, April, 1966.

J. L. Massey, "Note on Finite-Memory Sequential Machines," *IEEETEC*, Vol. EC-15, No. 4, pp. 658–659, Aug., 1966.

A. Grasselli and F. Luccio, "A Method for the Combined Row-Column Reduction of Flow Tables," *IEEE Conference Record of 1966 Seventh Annual Symposium on Switching and Automata Theory*, 16 C 40, pp. 136–147, Oct., 1966.

F. Luccio, "Reduction of the Number of Columns in Flow Table Minimization," *IEEETEC*, Vol. EC-15, No. 5, pp. 803–805, Oct., 1966.

W. Meusel, "A Note on Internal State Minimization in Incompletely Specified Sequential Networks," *IEEETEC*, Vol. EC-16, No. 4, pp. 508–509, Aug., 1967.

A. Bouchet, "An Algebraic Method for Minimizing the Number of States in an Incomplete Sequential Machine," *IEEETC*, Vol. C-17, No. 8, pp. 795–798, Aug., 1968.

A. K. Choudhury, A. K. Basu, and S. C. DeSarkar, "On the Determination of the Maximum Compatibility Classes," *IEEETC*, Vol. C-18, No. 7, p. 665, July, 1969.

J. Kella, "State Minimization of Incompletely Specified Sequential Machines," *IEEETC*, Vol. C-19, No. 4, pp. 342–348, April, 1970.

F. M. Brown, "Comment on 'The Determination of the Maximum Compatibility Classes,' " A. K. Choudhury, A. K. Basu, and S. C. DeSarkar, "Authors' Reply," *IEEETC*, Vol. C-19, No. 5, p. 459, May, 1970.

R. G. Bennetts, "An Improved Method of Prime C-Class Derivation in the State Reduction of Sequential Networks," *IEEETC*, Vol. C-20, No. 2, pp. 229–231, Feb., 1971.

J. KELLA, "Sequential Machine Identification," *IEEETC*, Vol. C-20, No. 3, pp. 332–338, March, 1971.

H. A. CURTIS, "The Further Reduction of CC-Tables," *IEEETC*, Vol. C-20, No. 4, pp. 454–456, April, 1971.

D. PAGER, "Conditions for the Existence of Minimal Closed Covers Composed of Maximal Compatibles," *IEEETC*, Vol. C-20, No. 4, pp. 450–452, April, 1971.

P. K. S. ROY and C. L. SHENG, "A Decomposition Method of Determining Maximum Compatibles," *IEEETC*, Vol. C-21, No. 3, pp. 309–312, March, 1972.

H.-D. EHRICH, "A Note on State Minimization of a Special Class of Incomplete Sequential Machines," *IEEETC*, Vol. C-21, No. 5, pp. 500–502, May, 1972.

I. TOMESCU, "A Matrix Method for Determining All Pairs of Compatible States of a Sequential Machine," *IEEETC*, Vol. C-21, No. 5, pp. 502–503, May, 1972.

G. H. WILLIAMS and E. G. ZAVISCA, "Comments on 'Sequential Machine Identification,'" *IEEETC*, Vol. C-21, No. 6, p. 616, June, 1972.

J. KIM and M. M. NEWBORN, "The Simplification of Sequential Machines with Input Restrictions," *IEEETC*, Vol. C-21, No. 12, pp. 1440–1443, Dec., 1972.

S. R. DAS, "On a New Approach for Finding All the Modified Cut-Sets in an Incompatibility Graph," *IEEETC*, Vol. C-22, No. 2, pp. 187–193, Feb., 1973.

P. K. GUPTA and D. L. DIETMEYER, "Fast State Minimization of Incompletely Specified Sequential Machines," *IEEETC*, Vol. C-22, No. 2, pp. 215–217, Feb., 1973.

SURESHCHANDER, "Comments on 'Decomposition Method of Determining Maximum Compatibles,'" *IEEETC*, Vol. C-22, No. 6, p. 627, June, 1973.

P. K. S. ROY and C. L. SHENG, "Authors' Reply," *IEEETC*, Vol. C-22, No. 6, p. 627, June, 1973.

C. P. PFLEEGER, "State Reduction in Incompletely Specified Finite-State Machines," *IEEETC*, Vol. C-22, No. 12, pp. 1099–1102, Dec., 1973.

C.-C. YANG, "Closure Partition Method for Minimizing Incomplete Sequential Machines," *IEEETC*, Vol. C-22, No. 12, pp. 1109–1122, Dec., 1973.

Chapter 11

D. A. HUFFMAN, "A Study of the Memory Requirements of Sequential Switching Circuits," *Research Lab. of Electronics Technical Report 293*, M.I.T., March 14, 1955.

E. J. MCCLUSKEY, JR. and S. H. UNGER, "A Note on the Number of Internal Assignments for Sequential Switching Circuits," *IRETEC*, Vol. EC-8, No. 4, pp. 439–440, Dec., 1959.

R. BIANCHINI and C. FREIMAN, "On Internal Variable Assignments for Sequential Switching Circuits," *IRETEC*, Vol. EC-10, No. 1, pp. 95–96, March, 1961.

J. HARTMANIS, "On the State Assignment Problem for Sequential Machines. I," *IRETEC*, Vol. EC-10, No. 2, pp. 157–165, June, 1961.

W. H. DAVIDOW, "A State Assignment Technique for Synchronous Sequential Networks," *Stanford Electronics Laboratories Technical Report* 1901–1, Stanford University, July 20, 1961.

R. E. STEARNS and J. HARTMANIS, "On the State Assignment Problem for Sequential Machines II," *IRETEC*, Vol. EC-10, No. 4, pp. 593–603, Dec., 1961.

D. B. ARMSTRONG, "A Programmed Algorithm for Assigning Internal Codes to Sequential Machines," *IRETEC*, Vol. EC-11, No. 4, pp. 466–472, Aug., 1962.

D. B. ARMSTRONG, "On the Efficient Assignment of Internal Codes to Sequential Machines," *IRETEC*, Vol. EC-11, No. 5, pp. 611–622, Oct., 1962.

A. J. NICHOLS, "Comments on Armstrong's State Assignment Techniques," *IEEETEC*, Vol. EC-12, No. 4, pp. 407–409, Aug. 1963.

Z. KOHAVI, "Secondary State Assignment for Sequential Machines," *IEEETEC*, Vol. EC-13, No. 3, pp. 193–203, June, 1964.

T. A. DOLOTTA and E. J. MCCLUSKEY, JR., "The Coding of Internal States of Sequential Circuits," *IEEETEC*, Vol. EC-13, No. 5, pp. 549–562, Oct, 1964.

R. M. KARP, "Some Techniques of State Assignment for Synchronous Sequential Machines," *IEEETEC*, Vol. EC-13, No. 5, pp. 507–518, Oct., 1964.

R. M. KARP, "Correction to 'Some Techniques of State Assignment for Synchronous Sequential Machines,'" *IEEETEC*, Vol. EC-14, No. 1, p. 61, Feb., 1965.

K. E. BATCHER, "On the Number of Stable States in a NOR Network," *IEEETEC*, Vol. EC-14, No. 6, pp. 931–932, Dec., 1965.

F. M. BROWN, "Code Transformation in Sequential Machines," *IEEETEC*, Vol. EC-14, No. 6, pp. 822–829, Dec., 1965.

J. HARTMANIS, "Two Tests for the Linearity of Sequential Machines," *IEEETEC*, Vol. EC-14, No. 6, pp. 781–786, Dec., 1965.

Z. KOHAVI, "Reduction of Output Dependency in Sequential Machines," *IEEETEC*, Vol. EC-14, No. 6, pp. 932–934, Dec., 1965.

H. FRANK and S. S. YAU, "Improving Reliability of a Sequential Machine by Error-Correcting State Assignments," *IEEETEC*, Col. EC-15, No. 1, pp. 111–113, Feb., 1966.

T. U. ZAHLE, "On Coding the States of Sequential Machines with the Use of Partition Pairs," *IEEETEC*, Vol. EC-15, No. 2, pp. 249–253, April, 1966.

J. H. TRACEY, "Internal State Assignments for Asynchronous Sequential Machines," *IEEETEC*, Vol. EC-15, No. 4, pp. 551–560, Aug., 1966.

W. A. DAVIS, "An Approach to the Assignment of Input Codes," *IEEETEC*, Vol. EC-16, No. 4, pp. 435–442, Aug., 1967.

D. R. HARING and A. K. SUSSKIND, "Realizations of Synchronous Sequential Machines," *IEEETEC*, Vol. EC-16, No. 5, pp. 686–687, Oct., 1967.

G. G. LANGDON, JR., "Delay-Free Asynchronous Circuits with Constrained Line Delays," *IEEETC*, Vol. C-18, No. 2, pp. 175–181, Feb., 1969.

P. T. HULINA, "On Tracey's Internal State Assignment Method," *IEEETC*, Vol. C-19, No. 5, p. 458, May, 1970.

G. K. MAKI and J. H. TRACEY, "State Assignment Selection in Asynchronous Sequential Circuits," *IEEETC*, Vol. C-19, No. 7, pp. 641–644, July, 1970.

D. P. BURTON, "Comment on 'Delay-Free Asynchronous Circuits with Constrained Line Delays,'" *IEEETC*, Vol. C-19, No. 10, p. 982, Oct., 1970.

J. G. BREDESON and P.T. HULINA, "Generation of a Clock Pulse for Asynchronous Sequential Machines to Eliminate Critical Races," *IEEETC*, Vol. C-20, No. 2, pp. 255–256, Feb., 1971.

G. MAGÓ, "Realization Methods for Asynchronous Sequential Circuits," *IEEETC*, Vol. C-20, No. 3, pp. 290–297, March, 1971.

L. L. KINNEY, "A Characterization of Some Asynchronous Sequential Networks and State Assignments," *IEEETC*, Vol. C-20, No. 4, pp. 426–436, April, 1971.

G. K. MAKI and J. H. TRACEY, "A State Assignment Procedure for Asynchronous Sequential Circuits," *IEEETC*, Vol. C-20, No. 6, pp. 666–668, June, 1971.

G. SAUCIER, "State Assignment of Asynchronous Sequential Machines Using Graph Techniques," *IEEETC*, Vol. C-21, No. 3, pp. 282–288, March, 1972.

R. PARCHMANN, "The Number of State Assignments for Sequential Machines," *IEEETC*, Vol. C-21, No. 6, pp. 613–614, June, 1972.

G. K. MAKI, D. H. SAWIN, III, and B.-R. A. JENG, "Improved State Assignment Selection Tests," *IEEETC*, Vol. C-21, No. 12, pp. 1443–1444, Dec., 1972.

J. R. STORY, H. J. HARRISON, and E. A. REINHARD, "Optimum State Assignment for Synchronous Sequential Circuits," *IEEETC*, Vol. C-21, No. 12, pp. 1365–1373, Dec., 1972.

Chapter 12

P. J. GRAHAM and R. J. DISTLER, "RST Flip-Flop Input Equations," *IEEETEC*, Vol. EC-16, No. 4, pp. 443–445, Aug., 1967.

SURESHCHANDER, "Comments on 'RST Flip-Flop Input Equations,'" *IEEETC*, Vol. C-17, No. 7, pp. 701–702, July, 1968.

M. D. MCINTOSH and B. L. WEINBERG, "On Asynchronous Machines with Flip Flops," *IEEETC*, Vol. C-18, No. 5, p. 473, May, 1969.

M. P. MARCUS, "S-R-T Flip-Flop," *IEEETC*, Vol. C-18, No. 6, pp. 568–569, June, 1969.

H. A. CURTIS, "Systematic Procedures for Realizing Synchronous Sequential Machines Using Flip-Flop Memory: Part I," *IEEETC*, Vol. C-18, No. 12, pp. 1121–1127, Dec., 1969.

H. A. CURTIS, "Systematic Procedures for Realizing Synchronous Sequential Machines Using Flip-Flop Memory: Part II," *IEEETC*, Vol. C-19, No. 1, pp. 66–73, Jan., 1970.

SURESHCHANDER, "RST Flip-Flop Input Equations," *IEEETC*, Vol. C-19, No. 11, p. 1118, Nov., 1970.

H. A. CURTIS, "The Realization of Polylinear Sequential Circuits Using Flip-Flop Memory," *IEEETC*, Vol. C-20, No. 1, pp. 87–94, Jan., 1971.

F. M. Brown, "Single-Parameter Solutions for Flip-Flop Equations," *IEEETC*, Vol. C-20, No. 4, pp. 452–454, April, 1971.

J. A. Brzozowski, "About Feedback and SR Flip-Flops," *IEEETC*, Vol. C-20, No. 4, p. 476, April, 1971.

H. Y. H. Chuang and S. Das, "Synthesis of Multiple-Input Change Asynchronous Machines Using Controlled Excitation and Flip-Flops," *IEEETC*, Vol. C-22, No. 12, pp. 1103–1109, Dec., 1973.

H. T. Hao and M. M. Newborn, "A Study of Trigger Machines," *IEEETC*, Vol. C-22, No. 12, pp. 1123–1131, Dec., 1973.

Chapter 13

D. A. Huffman, "The Design and Use of Hazard-Free Switching Networks," *JACM*, Vol. 4, No. 1, pp. 47–62, Jan., 1957.

M. Kliman and O. Lowenschuss, "Asynchronous Electronic Switching Circuits," *IRE National Conventional Record*, Part 4, pp. 267–274, 1959.

S. H. Unger, "Hazards and Delays in Asynchronous Sequential Switching Circuits," *IRETCT*, Vol. CT-6, No. 1, pp. 12–25, March, 1959.

D. E. Muller, "Treatment of Transition Signals in Electronic Switching Circuits by Algebraic Methods," *IRETEC*, Vol. EC-8, No. 3, p. 401, Sept., 1959.

M. P. Marcus, "Relay Essential Hazards," *IEEETEC*, Vol. EC-12, No. 4, pp. 405–407, Aug., 1963.

M. Yoeli and S. Rinon, "Application of Ternary Algebra to the Study of Static Hazards," *JACM*, Vol. 11, No. 1, pp. 84–97, Jan., 1964.

E. B. Eichelberger, "Hazard Detection in Combinational and Sequential Switching Circuits," *IBM Journal of Research and Development*, Vol. 9, No. 2, pp. 90–99, March, 1965.

E. J. Smith and Z. Kohavi, "Synthesis of Multiple Sequential Machines," *IEEE Conference Record of 1966 Seventh Annual Symposium on Switching and Automata Theory*, 16 C 40, pp. 160–171, Oct., 1966.

D. B. Armstrong, A. D. Friedman and P. R. Menon, "Realization of Asynchronous Sequential Circuits Without Inserted Delay Elements," *IEEETC*, Vol. C-17, No. 2, pp. 129–134, Feb., 1968.

S. H. Unger, "A Row Assignment for Delay-Free Realizations of Flow Tables Without Essential Hazards," *IEEETC*, Vol. C-17, No. 2, pp. 146–151, Feb., 1968.

W. S. Meisel and R. S. Kashef, "Hazards in Asynchronous Sequential Circuits," *IEEETC*, Vol. C-18, No. 8, pp. 752–759, Aug., 1969.

J. Hlavička, "Essential Hazard Correction Without the Use of Delay Elements," *IEEETC*, Vol. C-19, No. 3, pp. 232–238, March, 1970.

R. R. HACKBART and D. L. DIETMEYER, "The Avoidance and Elimination of Function Hazards in Asynchronous Sequential Circuits," *IEEETC*, Vol. C-20, No. 2, pp. 184–189, Feb., 1971.

J. G. BREDESON and P. T. HULINA, "Elimination of Static and Dynamic Hazards for Multiple Input Changes in Combinational Switching Circuits," *Information and Control*, Vol. 20, No. 2, pp. 114–124, March, 1972.

M. SERVÍT, "Hazard Correction in Asynchronous Sequential Circuits Using Inertial Delay Elements," *IEEETC*, Vol. C-22, No. 11, pp. 1041–2, Nov., 1973.

Appendix A

J. RIORDAN and C. E. SHANNON, "The Number of Two-Terminal Series-Parallel Networks," *Journal of Mathematics and Physics*, Vol. 21, No. 2, pp. 83–93, 1942.

G. A. MONTGOMERIE, "Sketch for an Algebra of Relay and Contactor Circuits," *J. IEE*, Vol. 95, No. 36, pp. 303–312, July, 1948.

W. KEISTER, "The Logic of Relay Circuits," *AIEE Transactions*, Vol. 68, pp. 571–576, 1949.

C. E. SHANNON, "The Synthesis of Two-Terminal Switching Circuits," *BSTJ*, Vol. 28, No. 1, pp. 59–98, Jan., 1949.

D. R. BROWN and N. ROCHESTER, "Rectifier Networks for Multiposition Switching," *Proc. IRE*, Vol. 37, No. 2, pp. 139–147, Feb., 1949.

C. E. SHANNON and E. F. MOORE, "Machine Aid for Switching Circuit Design," *Proc. IRE*, Vol. 41, No. 10, pp. 1348–1351, Oct., 1953.

F. E. HOHN and L. R. SCHISSLER, "Boolean Matrices and the Design of Combinational Relay Switching Circuits," *BSTJ*, Vol. 34, No. 1, pp. 177–202, Jan., 1955.

B. D. RUDIN, "A Theorem on SPDT Switching Circuits," *Proc. of the Western Joint Computer Conference*, pp. 129–132, March 1–3, 1955. (Published by the *IRE*.)

F. E. HOHN, "A Matrix Method for the Design of Relay Circuits," *IRETCT* Vol. CT-2, No. 2, pp. 154–161, June, 1955.

A. W. BURKS, et al., "The Folded Tree," *Journal of the Franklin Institute*, Vol. 260, Part I, No. 1, pp. 9–24, July, 1955; Part II, No. 2, pp. 115–126, Aug., 1955.

A. H. SCHEINMAN, "A Numerical-Graphical Method for Synthesizing Switching Circuits," *AIEE Transactions, Part I, Communication and Electronics*, pp. 687–689, 1957.

P. CALINGAERT, "Multiple-Output Relay Switching Circuits," *Proceedings of an International Symposium on the Theory of Switching, Part II*, Harvard University, Cambridge, Mass., pp. 59–73, April, 1957.

F. E. HOHN, "2N-Terminal Contact Networks," *Proceedings of an International Symposium on the Theory of Switching, Part II*, Harvard University, Cambridge, Mass., pp. 51–58, April, 1957.

G. N. Povarov, "A Mathematical Theory for the Synthesis of Contact Networks with One Input and *k* Outputs," *Proceedings of an International Symposium on the Theory of Switching, Part II*, Harvard University, Cambridge, Mass., pp. 74–94, April, 1957.

V. N. Roginskij, "A Graphical Method for the Synthesis of Multiterminal Contact Networks," *Proceedings of an International Symposium on the Theory of Switching, Part II*, Harvard University, Cambridge, Mass., pp. 302–315, April, 1957.

W. Semon, "Matrix Methods in the Theory of Switching," *Proceedings of an International Symposium of the Theory of Switching, Part II*, Harvard University, Cambridge, Mass., pp. 13–50, April, 1957.

M. P. Marcus, "Minimization of the Partially-Developed Transfer Tree," *IRETEC*, Vol. EC-6, No. 2, pp. 92–95, June, 1957.

W. Semon, "Synthesis of Series-Parallel Network Switching Functions," *BSTJ*, Vol. 37, No. 4, pp. 877–898, July, 1958.

R. E. Miller, "Formal Analysis and Synthesis of Bilateral Switching Networks," *IRETEC*, Vol. EC-7, No. 3, pp. 231–244, Sept., 1958.

A. H. Scheinman, "The Numerical-Graphical Method in the Design of Multi-terminal Switching Circuits," *AIEE Transactions, Part I, Communication and Electronics*, Vol. 78, pp. 515–519, Nov., 1959.

E. L. Lawler and G. A. Salton, "The Use of Parenthesis-Free Notation for the Automatic Design of Switching Circuits," *IRETEC*, Vol. EC-9, No. 3, pp. 342–352, Sept., 1960.

E. F. Moore, "Minimal Complete Relay Decoding Networks," *IBM Journal of Research and Development*, Vol. 4, No. 5, pp. 525–531, Nov., 1960.

R. A. Short, "The Design of Complementary-Output Networks," *IRETEC*, Vol. EC-11, No. 6, pp. 743–753, Dec., 1962.

R. A. Short, "Correction to 'The Design of Complementary-Output Networks,'" *IEEEETEC*, Vol. EC-12, No. 3, p. 232, June, 1963.

E. L. Lawler, "The Minimal Synthesis of Tree Structures," *Proc. of the Fourth Annual Symposium on Switching Circuit Theory and Logical Design*, S-156, pp. 63–82, Sept., 1963. (Published by the *IEEE*.)

Appendix B

C. E. Shannon, "A Symbolic Analysis of Relay and Switching Circuits," *Trans. AIEE*, Vol. 57, pp. 713–723, 1938.

S. H. Washburn, "Relay 'Trees' and Symmetric Circuits," *Trans. AIEE*, Part I, Vol. 68, pp. 582–586, 1949.

S. H. Caldwell, "The Recognition and Identification of Symmetric Switching Functions," *Trans. AIEE*, Part II, Vol. 73, pp. 142–147, May, 1954.

E. J. McCluskey, Jr., "Detection of Group Invariance or Total Symmetry of a Boolean Function" *BSTJ*, Vol. 35, No. 6, pp. 1445–1453, Nov., 1956.

M. P. Marcus, "The Detection and Identification of Symmetric Switching Functions with the Use of Tables of Combinations," *IRETEC*, Vol. EC-5, No. 4, pp. 237–239, Dec., 1956.

G. Epstein, "Synthesis of Electronic Circuits for Symmetric Functions," *IRETEC*, Vol. EC-7, No. 1, pp. 57–60, March, 1958.

B. Elspas, "Self-Complementary Symmetry Types of Boolean Functions," *IRETEC*, Vol. EC-9, No. 2, pp. 264–266, June, 1960.

R. F. Arnold and M. A. Harrison, "Algebraic Properties of Symmetric and Partially Symmetric Boolean Functions," *IEEETEC*, Vol. EC-12, No. 3, pp. 244–251, June, 1963.

A. Mukhopadhyay, "Detection of Total or Partial Symmetry of a Switching Function with the Use of Decomposition Charts," *IEEETEC*, Vol. EC-12, No. 5, pp. 553–557, Oct., 1963.

C. L. Sheng, "Detection of Totally Symmetric Boolean Functions," *IEEETEC*, Vol. EC-14, No. 6, pp. 924–926, Dec., 1965.

A. K. Choudhury, S. R. Das, and C. L. Sheng, "Comment on 'Detection of Totally Symmetric Boolean Functions,' " *IEEETEC*, Vol. EC-15, No. 5, p. 813, Oct., 1966.

S. Even, I. Kohavi, and A. Paz, "On Minimal Modulo 2 Sums of Products for Switching Functions," *IEEE Conference Record of 1966 Seventh Annual Symposium on Switching and Automata Theory*, 16 C 40, pp. 201–206, Oct., 1966.

A. Mukhopadhyay, "Symmetric Ternary Switching Functions," *IEEETEC*, Vol. EC-15, No. 5, pp. 731–739, Oct., 1966.

S. Even and I. Kohari, "On Minimal Modulo 2 Sums of Products for Switching Functions," *IEEETEC*, Vol. EC-16, No. 5, pp. 671–674, Oct., 1967.

D. L. Dietmeyer and P. R. Schneider, "Identification of Symmetry, Redundancy and Equivalence of Boolean Functions," *IEEETEC*, Vol. EC-16, No. 6, pp. 804–817, Dec., 1967.

R. C. Born and A. K. Scidmore, "Transformation of Switching Functions to Completely Symmetric Switching Functions," *IEEETC*, Vol. C-17, No. 6, pp. 596–599, June, 1968.

N. N. Biswas, "On Identification of Totally Symmetric Boolean Functions," *IEEETC*, Vol. C-19, No. 7, pp. 645–648, July, 1970.

S. R. Das and C. L. Sheng, "On Detecting Total or Partial Symmetry of Switching Functions," *IEEETC*, Vol. C-20, No. 3, pp. 352–355, March, 1971.

S. S. Yau and Y. S. Tang, "Transformation of an Arbitrary Switching Function to a Totally Symmetric Function," *IEEETC*, Vol. C-20, No. 12, pp. 1606–1609, Dec., 1971.

S. S. Yau and Y. S. Tang, "On Identification of Redundancy and Symmetry of Switching Functions," *IEEETC*, Vol. C-20, No. 12, pp. 1609–1613, Dec., 1971.

R. C. BORN, "An Iterative Technique for Determining the Minimal Number of Variables for a Totally Symmetric Function with Repeated Variables," *IEEETC*, Vol. C-21, No. 10, pp. 1129–1131, Oct., 1972.

B. DAHLBERG, "On Symmetric Functions with Redundant Variables—Weighted Functions," *IEEETC*, Vol. C-22, No. 5, pp. 450–458, May, 1973.

N. S. KHABRA and S. R. DAS, "Multiform Partial Symmetry and Linearity," *IEEETC*, Vol. C-22, No. 8, p. 804, Aug., 1973.

Appendix C

A. E. RITCHIE, "Sequential Aspects of Relay Circuits," *AIEE Transactions*, Part I, Vol. 68, pp. 577–581, 1949.

W. S. BENNETT, "Minimizing and Mapping Sequential Circuits," *AIEE Communication and Electronics*, pp. 443–447, Sept., 1955.

J. A. BROZOZOWSKI, "Some Problems in Relay Circuit Design," *IEEETEC*, Vol. EC-14, No. 4, pp. 630–634, Aug., 1965.

Answers and Solutions to Problems

Chapter 1

1. (a) 1
   (b) 0
   (c) 1
   (d) $\bar{C}$
   (e) $\bar{C}$
   (f) 0

2. (a) $[(A + \bar{B})C + \bar{D}]E + \bar{F}$
   (b) $[\bar{S} + W(\bar{I} + \bar{T}C)]\bar{H}$

3. (a) $(A + E)C(DF + B)$
   (b) $BF + E + (A + C)D$
   (c) $B(D + E)[AC + F(G + H)]$
   (d) $CE + F + (A + B)(D + GH)$
   (e) $A(C + D)[B(\bar{E} + F) + \bar{G}\bar{H}]$
   (f) $\bar{C} + EF + (A\bar{B} + D)(\bar{G} + \bar{H}K)$

275

4. (a) $A\bar{C}$

(b) $A\bar{C} + A\bar{B} + \bar{C}D$

(c) $(\bar{A} + B)(B + CD) = B + \bar{A}CD$

(d) $\bar{H}E + DE + G\bar{H} + HF\bar{E}$

(e) $(L + \bar{P})(L + M)(Q + P + \bar{L})(\bar{P} + N)$

(f) $(A + BC)(A + D)(BC + E)(\bar{A} + \bar{B} + \bar{C} + F)$

5. (a) $AB + \bar{B}C\bar{D} + C\bar{D}E$

(b) $B\bar{C}\bar{E} + AE$

(c) $(A + B)(C + \bar{D})$

(d) $(B + \bar{C} + \bar{D})(A + D)$

(e) $AD + B\bar{C}$

(f) $\bar{A}B\bar{C} + CD$

(g) $(\bar{A} + B + \bar{C})(C + D)(\bar{A} + B + E)$

(h) $(\bar{A} + B + \bar{C})(C + D)$

6. (a) $AC + \bar{B}\bar{A} + \bar{D}\bar{B} + \bar{C}EB + \bar{G}C$

(b) $(P + I)(\bar{I} + \bar{T})(P + A)(\bar{P} + O + T)(U + \bar{T})$

(c) $CE + \bar{D}\bar{E} + BC$

(d) $(\bar{A} + B)(C + A)(B + D)$

(e) $AF + \bar{E}\bar{F} + A\bar{B} + AD$

(f) $\bar{K}L + \bar{L}M + HM + \bar{G}\bar{M}$

(g) $X\bar{Y} + \bar{X}Z + \bar{Y}Z + X\bar{Z} = X\bar{Y} + \bar{X}Z + X\bar{Z}$ or $\bar{X}Z + \bar{Y}Z + X\bar{Z}$

(h) $(\bar{A} + B)(A + \bar{C})(B + \bar{C})(C + \bar{A}) = (\bar{A} + B)(A + \bar{C})(C + \bar{A})$ or $(A + \bar{C})(B + \bar{C})(C + \bar{A})$

7. (a) $(A + \bar{D}E)(\bar{A} + B + \bar{C})$

(b) $[\bar{D} + E(F + G)](D + A + \bar{B}\bar{C})$

8. (a) $A(\bar{D} + \bar{E}) + \bar{A}BC$

(b) $\bar{D}(E + F)G + D(A\bar{B} + \bar{C})$

9. (a) $(A + BC + D + E)(\bar{A} + G + C + FE)$
or $A(G + C + FE) + \bar{A}(BC + D + E)$

(b) $[A + \bar{B}(\bar{D} + E)(G + \bar{H} + J)][\bar{A} + C(\bar{D} + E + F)(G + \bar{H})]$
or $AC(\bar{D} + E + F)(G + \bar{H}) + \bar{A}\bar{B}(\bar{D} + E)(G + \bar{H} + J)$

10. $\bar{A}C + \bar{A}B + A\bar{C} + A\bar{B} + \bar{B}C + B\bar{C} =$
$\bar{A}B + \bar{B}C + A\bar{C}$

$(\bar{A} + \bar{C})(A + B) + \bar{B}C$

$(\bar{B} + \bar{A})(B + C) + A\bar{C}$

$(\bar{C} + \bar{B})(C + A) + \bar{A}B$

$A\bar{B} + B\bar{C} + \bar{A}C$

$(A + C)(\bar{A} + \bar{B}) + B\bar{C}$

$(B + A)(\bar{B} + \bar{C}) + \bar{A}C$

$(C + B)(\bar{C} + \bar{A}) + A\bar{B}$

$$\bar{A}(B + C) + A(\bar{B} + \bar{C})$$
$$\bar{B}(A + C) + B(\bar{A} + \bar{C})$$
$$\bar{C}(A + B) + C(\bar{A} + \bar{B})$$
$$(\bar{A} + \bar{B} + \bar{C})(A + B + C)$$

Chapter 2

1. (a) $(\bar{A} + B)(\bar{A} + \bar{C})$

(b) $\bar{A}\bar{B}\bar{C} + \bar{A}\bar{B}C + \bar{A}B\bar{C} + \bar{A}BC + AB\bar{C}$

(c) $(\bar{A} + B + \bar{C})(\bar{A} + B + C)(\bar{A} + \bar{B} + \bar{C})$

2. (a) $AB + A\bar{C} + D$

(b) $(A + D)(B + \bar{C} + D)$

(c) $AB\bar{C}\bar{D} + AB\bar{C}D + ABC\bar{D} + ABCD + A\bar{B}\bar{C}D + A\bar{B}C\bar{D} + \bar{A}B\bar{C}D$
$+ \bar{A}\bar{B}CD + \bar{A}B\bar{C}D + \bar{A}BCD + A\bar{B}CD$

(d) $(A + \bar{B} + \bar{C} + D)(A + \bar{B} + C + D)(A + B + \bar{C} + D)$
$(A + B + C + D)(\bar{A} + B + \bar{C} + D)$

Chapter 4

1. $\bar{A}\bar{B} + A\bar{C}D + AB\bar{D}$

2. $f(A, B, C, D) = \prod (4, 5, 6, 8, 11, 15) + \prod_\phi (0, 7, 10)$

3. $(A + \bar{B})(\bar{A} + \bar{C} + \bar{D})(B + C + D)$ or $(A + \bar{B})(\bar{A} + \bar{C} + \bar{D})$
$(\bar{A} + B + D)$

4. $\bar{A}\bar{C}\bar{D}, \bar{A}B\bar{C}, \bar{A}CD, \bar{A}BD, B\bar{C}D, AC\bar{D}, \bar{B}\bar{D}, \bar{B}C$

5. (a) $7 (UV + UWX + UXY + VWZ + VYZ + WXZ + XYZ)$

(b) $\bar{A}\bar{B}\bar{E} + \bar{A}\bar{C}E$

(c) $BE + D + C\bar{E}$

7. 1: $\bar{A}B\bar{C}\bar{D} + \bar{B}\bar{D} + BCD$

2: $\bar{A}BCD + \bar{A}\bar{C}D + \bar{B}\bar{D}$

3: $\bar{A}\bar{B}C + \bar{A}B\bar{C}\bar{D} + \bar{A}BCD$

8.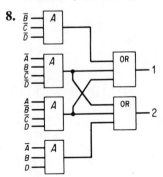

Chapter 5

1. $A + \bar{D} + \bar{B}C$

3. $B\bar{D} + \bar{C}D + \bar{A}D$ or $(B + D)(\bar{A} + \bar{C} + \bar{D})$

4. $\bar{A}B + \bar{C}D + B\bar{C} + \bar{A}D$ or $(\bar{A} + \bar{C})(B + D)$

6. $\bar{A}B\bar{C} + ABD\bar{E} + \bar{B}C\bar{D} + A\bar{B}\bar{D} + \bar{A}\bar{C}\bar{E}$

or $\bar{A}B\bar{C} + ABD\bar{E} + \bar{B}C\bar{D} + A\bar{B}\bar{D} + \bar{B}\bar{D}\bar{E}$

or $\bar{A}B\bar{C} + ABD\bar{E} + \bar{B}C\bar{D} + A\bar{B}\bar{D} + \bar{B}\bar{C}\bar{E}$

7. $(\bar{B} + D)(B + \bar{D})$ or $\begin{array}{c}(A + C + \bar{E} + D)(\bar{A} + \bar{E} + \bar{B})(A + \bar{C} + \bar{D}) \\ \text{or} \\ (A + C + \bar{E} + B)(\bar{A} + \bar{E} + \bar{D})(A + \bar{C} + \bar{B})\end{array}$

8. $\bar{C}\bar{D}E\bar{F} + C\bar{D}\bar{E}F + \bar{A}CD\bar{F} + \bar{A}\bar{C}D\bar{E} + A\bar{B}\bar{D} + \bar{B}DEF$

Chapter 6

1.
$$\begin{array}{rcl}
1 \times & 1 &= 1 \\
1 \times & 2 &= 2 \\
0 \times & 4 &= 0 \\
1 \times & 8 &= 8 \\
1 \times & 16 &= 16 \\
1 \times & 32 &= 32 \\
\hline
& & 59 \text{ (base 10)}
\end{array}$$

$$\begin{array}{r|cc}
3 & 59 & 2 \\
3 & 19 & 1 \\
3 & 6 & 0 \\
3 & 2 & 2 \\
& 0 &
\end{array} \quad \text{2012 (base 3)}$$

2.
$$\begin{array}{rcl}
1 \times & 1 &= 1 \\
0 \times & 7 &= 0 \\
6 \times & 49 &= 294 \\
2 \times & 343 &= 686 \\
\hline
& & 981 \text{ (base 10)}
\end{array}$$

$$\begin{array}{r|cc}
6 & 981 & 3 \\
6 & 163 & 1 \\
6 & 27 & 3 \\
6 & 4 & 4 \\
& 0 &
\end{array} \quad \text{4313 (base 6)}$$

4.
$$\begin{array}{r|cc}
2 & 13 & 1 \\
2 & 6 & 0 \\
2 & 3 & 1 \\
2 & 1 & 1 \\
& 0 &
\end{array}$$

$$\begin{array}{r l}
 & .8125 \\
 & \times 2 \\
\hline
1 & .6250 \\
 & \times 2 \\
\hline
1 & .2500 \\
 & \times 2 \\
\hline
0 & .5000 \\
 & \times 2 \\
\hline
1 & .0000
\end{array} \quad \text{1101.1101 (base 2)}$$

Chapter 7

1.

1	2	3	4	5	6	7	8	
C_1	C_2	8	C_4	4	2	1	P	
0	1	0	1	**0**	1	0	1	$\longrightarrow$ 2
1	**0**	0	0	0	1	1	1	$\longrightarrow$ 3
1	0	0	1	1	0	0	**1**	$\longrightarrow$ 4
0	1	0	0	1	1	1	0	$\longrightarrow$? (Double error, no correction made)*

*Note that this digit could be a 5, 6, or 7:

0	1	0	0	1	**0**	1	**1**	$\longrightarrow$ 5
1	1	0	0	1	1	**0**	0	$\longrightarrow$ 6
0	**0**	0	**1**	1	1	1	0	$\longrightarrow$ 7

Chapter 9

1.

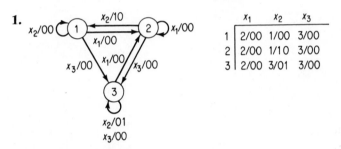

	x_1	x_2	x_3
1	2/00	1/00	3/00
2	2/00	1/10	3/00
3	2/00	3/01	3/00

2.

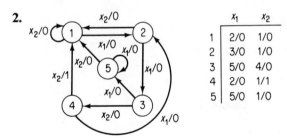

	x_1	x_2
1	2/0	1/0
2	3/0	1/0
3	5/0	4/0
4	2/0	1/1
5	5/0	1/0

3.

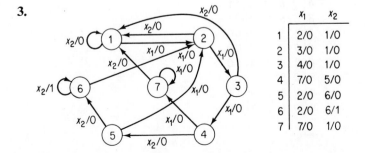

	x_1	x_2
1	2/0	1/0
2	3/0	1/0
3	4/0	1/0
4	7/0	5/0
5	2/0	6/0
6	2/0	6/1
7	7/0	1/0

5.

	$x_1 x_2$				
	00	01	11	10	Z
1	1	2	1	1	0
2	4	2	4	3	1
3	4	2	4	1	1
4	4	2	4	4	1

6. $x_1 x_2$

00	01	11	10	Z
①	4	5	7	0
②	4	6	8	1
2	③	5	8	0
1	④	6	8	1
1	3	⑤	8	0
2	4	⑥	7	1
1	4	6	⑦	0
2	3	6	⑧	1

7. $x_1 x_2$

00	01	11	10	Z
①	3	4	6	0
1	②	4	6	0
1	③	4	6	1
—	2	④	5	1
1	2	4	⑤	0
1	2	4	⑥	1

Chapter 10

1.

1-12 ≡ A
2-7 ≡ B
4-6 ≡ C
5-10 ≡ D

	x_1	x_2	x_3	x_4	$Z_1 Z_2$
A	D	B	11	A	1 1
B	3	B	11	A	0 1
3	3	B	C	A	0 0
C	D	B	C	A	0 0
D	D	B	C	A	1 0
8	D	8	C	A	0 1
9	D	8	C	9	1 1
11	D	8	11	9	0 0

2.

$x_1 x_2$

00	01	11	10	Z
(A)	B	C	D	0
(2)	B	7	D	0
A	(B)	C	D	1
A	(5)	C	12	1
A	5	(7)	D	0
A	B	(C)	D	0
A	B	C	(D)	1
2	B	C	(12)	0

1-3 ≡ A
4-6 ≡ B
8-9 ≡ C
10-11 ≡ D

4.

$x_1 x_2$

00	01	11	10	$Z_1 Z_2$
(A)	–	B	C	00
A	–	(B)	D	11
A	8	B	(C)	10
A	9	B	(D)	10
A	(8)	B	–	01
A	(9)	B	–	10

1-2 ≡ A
3-4 ≡ B
5-7 ≡ C
6-7 ≡ D

5. 16789, 124689, 356789, 45689

6.

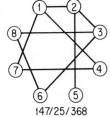

$x_1 x_2$

00	01	11	10
(1)	(4)	2	(7)
8	4	(2)	(5)
(8)	(3)	(6)	5

147/25/368

Chapter 11

1.

$x_1 x_2$

00	01	11	10	
(1)	2	5	(8)	a
(3)	4	(5)	7	b
3	(2)	(6)	7	c
1	(4)	5	(7)	d

Y_1

Y_2	0	1
0	a	d
1	c	b

$x_1 x_2$

$y_1 y_2$	00	01	11	10
00	(1)	2	5	(8)
01	3	(2)	(6)	7
11	(3)	4	(5)	7
10	1	(4)	5	(7)

$x_1 x_2$

$y_1 y_2$	00	01	11	10
00	00	01	10	00
01	11	01	01	11
11	11	10	11	10
10	00	10	11	10

3. One possible solution:

y_3 \ y_1y_2	00	01	11	10
0	a			
1	c	b		d

$y_1y_2y_3$ \ x_1x_2	00	01	11	10
000	①	②	3	④
001	1	6	③	10
011	⑤	⑥	3	⑦
010				
110				
111				
101	⑧	⑨	3	⑩
100				

$y_1y_2y_3$ \ x_1x_2	00	01	11	10
000	000	000	001	000
001	000	011	001	101
011	011	011	001	011
010	---	---	---	---
110	---	---	---	---
111	---	---	---	---
101	101	101	001	101
100	---	---	---	---

4. (a) a_3d_3

(b) $a_3a_1a_2d_1$

(c) $a_3a_1a_2d_3$

(d) $a_3a_1a_2d_2$ (may also terminate in d_1 or d_3)

(e) $a_3a_2d_1$ (may also terminate in d_3)

(f) $a_3a_2d_3$

(g) $a_3a_2d_2$ (may also terminate in d_1 or d_3)

In transitions (e), (f), and (g), a_1 must be directed to a_2.

5.

$y_1y_2y_3$ \ x_1x_2	00	01	11	10
000	000	010	001	100
001	001	000	101	000
011	---	---	---	010
010	110	010	010	010
110	110	111	010	110
111	101	111	111	011
101	001	101	101	111
100	---	---	---	110

Chapter 12

1.

y_1y_2	x_1	x_2	Z
00	10	00	0
10	01	11	1
01	01	00	0
11	10	11	1

Memory element 1:

x_1, y_1

y_2	0	1
0	1	0
1	0	1

x_2, y_1

y_2	0	1
0	0	1
1	0	1

Memory element 2:

y_1

y_2	0	1
0	0	1
1	1	0

y_1

y_2	0	1
0	0	1
1	0	1

3.

$S_1 = x_1 y_2 + x_2 \bar{y}_1 y_2$ $\qquad$ $S_2 = x_2 \bar{y}_1 \bar{y}_2$

$C_1 = x_1 \bar{y}_2 + x_2 y_1 y_2$ $\qquad$ $C_2 = x_2 \bar{y}_1 y_2$

$J_1 = x_1 y_2 + x_2 y_2$ $\qquad$ $J_2 = x_2 \bar{y}_1$

$K_1 = x_1 \bar{y}_2 + x_2 y_2$ $\qquad$ $K_2 = x_2 \bar{y}_1$

$T_1 = x_1 \bar{y}_1 y_2 + x_1 y_1 \bar{y}_2 + x_2 y_2$ $\qquad$ $T_2 = x_2 \bar{y}_1$

$S_1 = x_1 y_2$ $\qquad$ $S_2 = $ ——

$C_1 = x_1 \bar{y}_2$ $\qquad$ $C_2 = $ ——

$T_1 = x_2 y_2$ $\qquad$ $T_2 = x_2 \bar{y}_1$

$J_1 = x_1 y_2$ $\qquad$ $J_2 = $ —— $\qquad$ $J_2 = x_2 \bar{y}_1$

$K_1 = x_1 \bar{y}_2$ $\qquad$ $K_2 = $ —— or $K_2 = x_2 \bar{y}_1$

$T_1 = x_2 y_2$ $\qquad$ $T_2 = x_2 \bar{y}_1$ $\qquad$ $T_2 = $ ——

$D_1 = x_1 y_2 + x_2 \bar{y}_1 y_2 + x_2 y_1 \bar{y}_2$ $\qquad$ $D_2 = x_1 y_2 + x_2 \bar{y}_1 y_2 + x_2 y_1 y_2$

5. $Y_1 = x_1 x_2 \bar{y}_2 + \bar{x}_1 \bar{x}_2 y_2 + y_1 y_2$ or $(x_1 + y_2)(\bar{x}_1 + x_2 + y_1)(\bar{x}_2 + y_1 + \bar{y}_2)$

$Y_2 = x_1 \bar{x}_2 + \bar{y}_1 y_2 + x_1 y_1 + \bar{x}_2 y_1$ or $x_1 \bar{x}_2 + \bar{y}_1 y_2 + x_1 y_1 + \bar{x}_2 y_2$

or $(\bar{x}_2 + y_1 + y_2)(x_1 + \bar{x}_2 + \bar{y}_1)(x_1 + y_2)$

Note that the product of sums solutions, with $(x_1 + y_2)$ common to both Y_1 and Y_2, are optimum.

Chapter 13

1.

$x_1 x_2$

$x_1 y_2$	00	01	11	10
00	10	11	01	01
01	1-	11	0-	01
11	00	--	00	-0
10	-0	11	-1	10

Z map

$Z_1 = \bar{x}_1 \bar{y}_1 + y_1 \bar{y}_2$

or $(\bar{x}_1 + y_1)(\bar{y}_1 + \bar{y}_2)$

$Z_2 = x_1 \bar{y}_1 + x_2 \bar{y}_2 + \bar{y}_1 y_2$

or $x_1 \bar{y}_1 + x_2 \bar{y}_2 + \bar{x}_1 x_2$

or $x_1 \bar{y}_1 + x_2 \bar{y}_2 + x_2 \bar{y}_1$

or $(x_1 + x_2)(x_2 + \bar{y}_1)(\bar{y}_1 + \bar{y}_2)$

Note common $(\bar{y}_1 + \bar{y}_2)$ term in product of sums expressions.

3. (a)

	x_1	x_2
1	2/0	1/0
2	2/0	3/0
3	2/0	1/1

y_1y_2	x_1	x_2
00	01/0	00/0
01	01/0	11/0
11	01/0	00/1
10	– –	– –

Memory element 1

x_1
y_2 \ y_1	0	1
0	0	–
1	0	0

x_2
y_2 \ y_1	0	1
0	0	–
1	1	0

Memory element 2

y_1
y_2 \ y_1	0	1
0	1	–
1	1	1

y_1
y_2 \ y_1	0	1
0	0	–
1	1	0

S-C-T: $S_1 =$
 $C_1 = x_1$
 $T_1 = x_2 y_2$

S-C: $S_2 = x_1$
 $C_2 = x_2 y_1$

$Z = x_2 y_1$

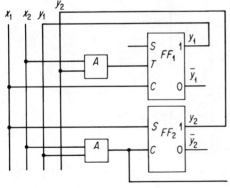

(a)

(b)

	x_1	x_2
1	2/0	1/0
2	2/0	3/0
3	2/0	1/1

$y_1 y_2$	x_1	x_2
00	01/0	00/0
01	01/0	10/0
10	01/0	00/1
11	– –	– –

Memory element 1

x_1

y_2 \ y_1	0	1
0	0	0
1	0	–

x_2

y_2 \ y_1	0	1
0	0	0
1	1	–

Memory element 2

y_2 \ y_1	0	1
0	1	1
1	1	–

y_2 \ y_1	0	1
0	0	0
1	0	–

$S\text{-}C\text{-}T:$ $S_1 = x_2 y_2$
      $C_1 = x_1$
      $T_1 = x_2 y_1$

$S\text{-}C:$ $S_2 = x_1$
     $C_2 = x_2$

$Z = x_2 y_1$

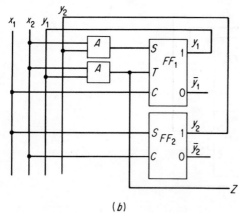

(b)

(c)

	x_1	x_2
1	2/0	1/0
2	2/0	3/0
3	2/0	1/1

$y_1 y_2$	x_1	x_2
00	11/0	00/0
11	11/0	01/0
01	11/0	00/1
10	– –	– –

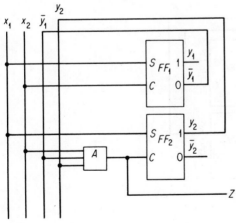

Memory element 1

$$x_1$$
y_2 \ y_1	0	1
0	1	–
1	1	1

$$x_2$$
y_2 \ y_1	0	1
0	0	–
1	0	0

Memory element 2

y_2 \ y_1	0	1
0	1	–
1	1	1

y_2 \ y_1	0	1
0	0	–
1	0	1

$S\text{-}C$: $S_1 = x_1$
$C_1 = x_2$

$S\text{-}C$: $S_2 = x_1$
$C_2 = x_2 \bar{y}_1$ or $x_2 \bar{y}_1 y_2$

$$Z = x_2 \bar{y}_1 y_2$$

(c)

4.

	x_1	x_2
1	2/0	1/0
2	3/0	1/0
3	3/0	1/1

$y_1 y_2$	x_1	x_2
00	01/0	00/0
01	11/0	00/0
11	11/0	00/1
10	--	--

Memory element 1

x_1

y_2 \ y_1	0	1
0	0	—
1	1	1

x_2

y_2 \ y_1	0	1
0	0	—
1	0	0

Memory element 2

y_1

y_2	0	1
0	1	—
1	1	1

y_1

y_2	0	1
0	0	—
1	0	0

$S\text{-}C$: $S_1 = x_1 y_2$
$C_1 = x_2$

$S\text{-}C$: $S_2 = x_1$
$C_2 = x_2$

$Z = x_2 y_1$

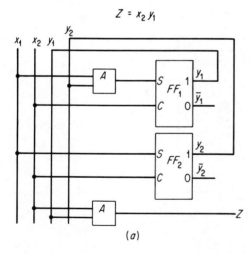

(a)

7.

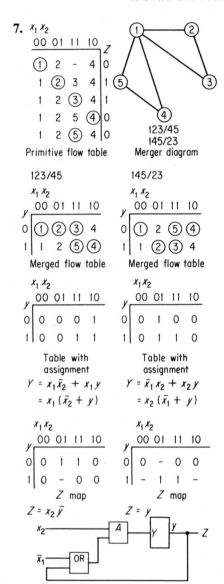

$x_1 x_2$

	00	01	11	10	Z
	①	2	–	4	0
	1	②	3	4	1
	1	2	③	4	1
	1	2	5	④	0
	1	2	⑤	4	0

Primitive flow table

123/45
145/23
Merger diagram

123/45

$x_1 x_2$

y

	00	01	11	10
0	①	②	③	4
1	1	2	⑤	④

Merged flow table

145/23

$x_1 x_2$

y

	00	01	11	10
0	①	2	⑤	④
1	1	②	③	4

Merged flow table

$x_1 x_2$

y

	00	01	11	10
0	0	0	0	1
1	0	0	1	1

Table with
assignment

$x_1 x_2$

y

	00	01	11	10
0	0	1	0	0
1	0	1	1	0

Table with
assignment

$Y = x_1 \bar{x}_2 + x_1 y$
$\quad = x_1 (\bar{x}_2 + y)$

$Y = \bar{x}_1 x_2 + x_2 y$
$\quad = x_2 (\bar{x}_1 + y)$

$x_1 x_2$

y

	00	01	11	10
0	0	1	1	0
1	0	–	0	0

Z map

$x_1 x_2$

y

	00	01	11	10
0	0	–	0	0
1	–	1	1	–

Z map

$Z = x_2 \bar{y}$

$Z = y$

x_2 ——— [A] —— [Y] — y — Z

$\bar{x}_1$ — [OR]

Index

Index